This Book is Dedicated

By Parul to:
My husband Amit Singh Minhas and my son Aarush Singh Minhas

By Prakash to:
My wife Jody Sampson and my mentor Edward E. Kirkbride

A NEW LANGUAGE OF SCHOOL Design

Evidence-Based Strategies for
Student Achievement & Well-being

EDITORS
Prakash Nair, AIA
Dr. Parul Minhas

Contributing Authors
Prakash Nair
Parul Minhas
Roni Zimmer Doctori
Louis Sirota
Bipin Bhadran
Gary Stager
Francesco Cupolo
Anna Harrison
Jeff Melendez
Paras Sareen
Karin Nakano

Association for Learning Environments

A NEW Language of School Design

$45 ARCHITECTURE, INNOVATIVE SCHOOL DESIGN, EDUCATION, SCHOOL REFORM

Copyright © Association for Learning Environments, Prakash Nair, Parul Minhas and Education Design International

ISBN 9798861102575

Cover & Book Design: Dmytro Zaporozhtsev

Photo Credits:
Education Design International
Prakash Nair
Parul Minhas
Louis Sirota
Fielding Nair International
Paul Brokering
Marissa Moss Photography
Andre Fanthome

Original Edition Printed in the United States of America

A NEW LANGUAGE OF SCHOOL DESIGN
Evidence-Based Strategies for Student Achievement & Well-being

TABLE OF CONTENTS

Foreword 1

Preface 3

Introduction
Why a New Language of School Design? 5

Chapter One - *Powered by AI*
Bold New Approaches to Education & Learning Places 9

Chapter Two - *Neuroarchitecture*
Health, Happiness, and Learning 19

Chapter Three - *Biophilic Design*
Learning Spaces Inspired by Nature 45

Chapter Four - *Salutogenic Design*
To Promote Student Health and Well-Being 85

Chapter Five - *Outdoor Learning*
Leave the Classroom Behind 125

Chapter Six - *Urban School Greening*
Stimulating Cognitive Growth and Fostering Environmental Stewardship 159

Chapter Seven - *Self-Determination Theory*
A Framework for School Design 165

Chapter Eight - *Unmasking Pseudoteaching*
Empowering Students Through Active and Authentic Learning 179

Chapter Nine - *Choice Architecture*
Nudging Behavioral Shifts for Holistic Well-being 191

Chapter Ten - *School Buildings*
The Last Domino 201

Chapter Eleven - *Design for Safety*
Balancing Physical and Psychological Considerations 203

Conclusion
Strategies for Successful Implementation 213

About the Authors 225

About A4LE 229

A NEW LANGUAGE OF SCHOOL DESIGN
Evidence-Based Strategies for Student Achievement & Well-being

FOREWORD

In our dynamic world, where almost every facet of life is undergoing rapid transformation, the realms of education and school architecture are no exception. Our schools not only need to respond to these changes but lead the way. A NEW LANGUAGE OF SCHOOL DESIGN, embodies this leadership, crystallizing our vision for the transformative journey of learning spaces.

As the CEO of A4LE, an association deeply rooted in the evolution of learning spaces, I've borne witness to many notable shifts in education over the years. While this is a significant milestone as our first commissioned book in our 100+ year history, it adds to our robust collection of publications. That is why we reached out to two better-known professionals in this area to compile their more recent works, including groundbreaking A4LE white papers, as the basis for this book. Our intent has always been to equip our membership with the knowledge and tools to stay at the forefront, and with this book, we bridge the gap between the present state of education and its promising future. We see this publication and the associated white papers as one of many member benefits to our association.

The core of this landmark publication revolves around the fusion of education and architecture, acknowledging that the environments we create for our learners play a profound role in shaping their experiences and, indeed, who they are as people. From insights into neuroarchitecture—understanding how learners interact with their surroundings on a cognitive level—to the holistic principles of biophilic design that integrates the natural world into learning spaces, this book is a definitive treatise on important discoveries that will help shape school design over the coming years. It underscores the imperative shift from passive to active learning environments, bolstered by designs that not only cater to academic needs but also the psychological and emotional well-being of students.

Adding depth to these architectural concepts, the inclusion of AI in education, Self-Determination Theory, and Choice Architecture reveals a tantalizing glimpse of the future. A future where each student, regardless of background or capability, can access a tailored, holistic, and globally connected educational experience.

At its core, this publication is a call to action offering practical solutions that can immediately be put into practice. Editor Prakash Nair, a preeminent architect with a vast global portfolio of innovative schools, brings a world-view approach. He has made extensive contributions in leading international journals and has authored several highly regarded books, including Learning by Design, and Blueprint for Tomorrow: Redesigning Schools for Student-Centered Learning published by Harvard Education Press. Alongside him, Dr. Parul Minhas, an internationally recognized author, speaker, and expert in her own right, offers a balanced perspective regarding the ways in which design intertwines with holistic health and well-being. These two prominent figures in our industry represent the excellent caliber of the membership of A4LE and we are proud to present their work as contributing members.

FOREWORD

As educators, policymakers, and architects read through these pages, it's essential to recognize the weight of our collective responsibility. As with all of the information we impart through our vast array of association content, the designs, methodologies, and strategies contained within this publication aren't just theoretical concepts; they present a clear roadmap for the next generation of learning spaces.

In introducing A NEW LANGUAGE OF SCHOOL DESIGN, I extend an invitation to all readers to embark on a journey. A journey to rethink, reimagine, and recreate spaces where every child not only learns but thrives.

John Ramsey

John Ramsey

CEO, A4LE

Dear Reader,

As you hold this book, you're not just holding another resource about school design. What you have in your hands is an invitation to be part of a transformative journey—a journey that redefines the language and landscape of educational spaces. This book is a playbook and a guide for those who aspire to design schools that matter, schools that acknowledge the needs of tomorrow's learners, and above all, schools that honor the human experience.

Our society is in the middle of an epochal transformation, steered by rapid technological advances, societal shifts, and a global awareness of the urgent challenges that confront us all. Despite the seismic changes in almost every domain of human activity, the architecture of schools and the pedagogical paradigms they embed have been stubbornly resistant to meaningful evolution.

It's precisely this resistance, often veiled as tradition, that we aim to challenge. The inertia in school design doesn't just stem from a reluctance to change or a fear of experimentation. It's also fueled by a disconnection from the rich tapestry of interdisciplinary research that could revolutionize the way we think about educational spaces. We've endeavored to bridge this gap by translating complex research from disciplines like neuroscience, psychology, health and well-being, and behavioral economics into a new language—one that serves as a call to action rather than mere academic rhetoric.

The terms and ideas we explore—educational architecture powered by AI, Neuroarchitecture, Biophilic Design, Salutogenic Design, and more—aren't exotic concepts relegated to specialized domains. They should be part of the basic vocabulary of all architects, educators, policymakers, and indeed anyone involved in the sphere of education.

Every chapter in this book is a deep dive into these critical components, offering you evidence-based insights, practical strategies, and inspirational examples from around the world. We present you with questions as well as answers, challenging your pre-existing notions while equipping you with the tools to build a future that our children not just deserve but sorely need.

We invite you to read, question, ponder, and most importantly, act. Whether you are an architect aiming to create impactful spaces, an educator passionate about your students' well-being, a policymaker with the power to drive change, or a parent concerned about your child's future, this book is for you. The future of our children and the generations to come hinges on our courage to speak a new language of school design contained in this book.

Thank you for embarking on this exciting journey with us.

Prakash Nair & Parul Minhas

INTRODUCTION

Why a New Language of School Design?

WHY A NEW LANGUAGE?

There is an essential new language of school design that this book introduces. While the terms themselves are not "new," it is our contention that in the context of school design, very few of the tens of thousands of architects who are currently designing learning spaces, the educational leaders to whom they report or even the students and staff who will occupy new and renovated learning environments are fully up to speed on the new language or its implications for both education and school design. But don't take our word for it. Go ahead and test yourself according to the criteria we have outlined below. Consider these terms and phrases:

1. Educational architecture powered by AI
2. Neuroarchitecture
3. Biophilic design
4. Salutogenic design
5. Pseudoteaching and its effects on learning spaces
6. Self-determination theory
7. Outdoor learning
8. Choice architecture

For each of the above, answer the following four questions. Only those who can answer in the affirmative to all four questions relative to each of the above eight terms are fully up to speed on the new language of school design. More importantly, these sub-set of the architectural/educational community have the best opportunity to create caring, nurturing places for children to grow and thrive.

Question One: Can you clearly define all eight of the above terms in no more than two sentences each?

Question Two: Can you describe how each of these terms directly applies as a critically important element in the context of modern school design?

Question Three: Have you been consciously and intentionally applying the principles associated with each of the seven terms in your own practice as a school design architect or as an educational leader or administrator responsible for the creation of new schools or the renovation of existing school buildings?

Question Four: To the extent that the principles associated with these new terminologies are being applied in your professional or administrative practice, how consciously are you able to help your school or schools transform their own culture to take full advantage of the opportunities afforded by the transformative designs?

Use the following chart *(Figure 1)* to score yourself before reading this book. The good news is that you will be able to answer YES to the first two questions across the eight topics covered by this book after you read it *(Figure 2)*. This will also set you well on your path to YES answers on the remaining two questions.

INTRODUCTION WHY A NEW LANGUAGE OF SCHOOL DESIGN?

Question	AI	Neuroarchitecture	Biophilic Design	Salutogenic Design	Pseudoteaching	SDT	Outdoor Learning	Choice Architecture
One								
Two								
Three								
Four								

FIGURE 1. Say "Yes" in any box across the eight domains where you can answer any of the four questions in the affirmative.

Question	AI	Neuroarchitecture	Biophilic Design	Salutogenic Design	Pseudoteaching	SDT	Outdoor Learning	Choice Architecture
One	Yes	Yes	Yes	Yes	Yes	Yes	Yes	Yes
Two	Yes	Yes	Yes	Yes	Yes	Yes	Yes	Yes
Three								
Four								

FIGURE 2. After reading this book, you will be able to answer YES to the first two questions across all eight topics.

CONTEXT

Even as the world evolves at breakneck speed around us, shockingly, the fundamental tenets of education and school building design across the globe remain unchanged. That means the vast majority of the 80 million students in the United States and hundreds of millions of children around the world spend the most formative years of their life preparing for a world that is far gone and in learning places that are not just dysfunctional but dehumanizing as well.

There are two complementary elements at work to maintain the status quo and both have largely withstood widespread reform efforts over the past few decades. At the physical level, we have classrooms as the basic building block of the school building and at the pedagogical level we have teacher direction at the core of curriculum delivery. In fact, that term "curriculum delivery" itself is obsolete because it should not be the role of schools to "deliver" a standardized, canned curriculum en masse to millions of children. There is overwhelming evidence that tomorrow's education should not be about consuming information but about autonomy, personal growth, critical thinking, teamwork, and creativity – all of which the majority of school buildings and the pedagogy they encourage militate against.

TOPICS COVERED

The premise of this book is that the design of school curricula and school buildings to support that curricula should be driven by established research about the impact of the learning environment on student health and well-being. To get there, we need a new 'language' of school design that explores ideas that have, until now, only been peripherally associated with the larger school building industry. Here are some of the important topics that we discuss in the book.

Powered by AI – Bold New Approaches to Education & Learning Places

AI is poised to revolutionize both schooling and the design of learning spaces, fundamentally transforming the landscape of education. With the integration of artificial intelligence in schools, personalized learning experiences will become the norm, catering to each student's unique needs and learning pace. AI-powered virtual tutors and intelligent learning platforms will provide instant feedback and tailored study plans, empowering students to excel and explore their interests. The design of physical learning spaces will adapt to accommodate interactive and immersive technologies, fostering collaborative and experiential learning. Collaborative AI platforms will facilitate global connections, fostering a diverse and inclusive educational experience where students from different backgrounds can collaborate and exchange ideas effortlessly. The future of education lies in AI's capacity to dissolve the boundaries of the traditional classroom and unlock a world of endless possibilities for learners to explore and grow.

Neuroarchitecture – Health, Happiness, and Learning

This chapter introduces the revolutionary concept of Neuroarchitecture, where the design of the built environment is rooted in neuroscience. This approach considers how different architectural elements affect brain function, emotion, and behavior, ultimately influencing students' health, happiness, and learning abilities. We explore the scientific basis of Neuroarchitecture and show how factors such as lighting, acoustics, spatial layout, color, and texture can enhance cognitive abilities, reduce stress, and foster a conducive atmosphere for effective learning and growth.

Biophilic Design – Learning Spaces Inspired by Nature

This is where we delve into the realm of Biophilic Design, a principle that strives to integrate nature into built environments. Drawing from a body of research, we'll discuss how this design approach taps into our intrinsic affinity for nature, promoting healthier, more creative, and more mindful learners. The biophilic approach to school design fosters an environment that not only motivates learning but also reduces stress, improves mood, and bolsters overall student well-being.

Salutogenic Design – To Promote Student Health and Well-Being

The focus of this chapter is on the Salutogenic Design principle, a model that prioritizes health and well-being in the design process. Salutogenic Design shifts the focus from disease prevention to health promotion. We'll discuss how this holistic approach can be applied to school design, creating spaces that actively promote physical, psychological, and social health. From air quality to ergonomics, we'll explore the many ways that Salutogenic Design can enhance student and staff well-being.

Outdoor Learning – Leave the Classroom Behind

This chapter presents the untapped potential of Outdoor Learning. As an innovative approach, it allows students to interact directly with the environment, fostering an appreciation for nature and the world around them. This chapter explores how designing school spaces for outdoor learning can enhance cognitive function, promote physical activity, and encourage cooperative learning. From outdoor classrooms to learning gardens, we'll look at the many ways that schools can break down the walls of the traditional classroom and bring learning into the great outdoors.

Self-Determination Theory – A Framework for School Design

Here, we integrate psychology into school design, drawing on Self-Determination Theory (SDT). SDT is a theory of motivation that focuses on individuals' inherent growth tendencies and innate psychological needs: autonomy, competence, and relatedness. We'll explore how school design can facilitate these needs, fostering a motivational climate that encourages engagement, curiosity, and positive behaviors. Whether it's through flexible learning spaces that empower students or collaboration zones that foster a sense of community, this chapter will shed light on the ways design can help schools nurture self-determined learners.

Choice Architecture

Choice architecture, a term coined by economist Richard H. Thaler and legal scholar Cass R. Sunstein in their book "Nudge: Improving Decisions about Health, Wealth, and Happiness," refers to the design of environments that subtly shape the way choices are made. This concept hinges on the premise that the ways choices are presented can significantly influence the decisions we make, often without our conscious awareness.

In the context of learning spaces, choice architecture provides a powerful tool for enhancing student engagement, autonomy, and overall well-being. As we navigate away from traditional classroom models, the design of learning spaces and the incorporation of choice architecture principles becomes increasingly important.

Overarching Theme

The overarching theme of this book is that wisdom in one field can be leveraged for success in another. For too long, the field of school architecture has existed in a silo divorced from the innovations happening all around it. It is our goal to bring creative ideas and innovative thinking from various other fields like Psychology, Neurology, Well-Being Research and Education and have them bear directly on the process for visioning, planning, designing, constructing, and using school facilities.

Roni Zimmer Doctori
Prakash Nair, AIA

Bold New Approaches to Education & Learning Places

Introduction

In a world constantly transforming, the once-clear demarcations between physical and digital spaces begin to blur. Our educational systems are at a pivotal juncture, balancing rich traditions with the beckoning of a digitally augmented future, powered by Artificial Intelligence (AI) and immersive technologies like Virtual Reality (VR) and Augmented Reality (AR).

FIGURE 3. We are at a pivotal juncture in our history, with a need to balance rich traditions with the beckoning of a digitally augmented future.

This chapter serves as an invitation—a beckoning into a journey that unveils the multifaceted tapestry of modern education. As we navigate this path, we'll explore the dynamic landscapes of learning environments, unravel the evolving role of educators, and confront the profound moral and ethical quandaries introduced by these innovations.

In this chapter, we talk about:

The Nature of Learning in AI-powered Environments: A deep dive into how AI reshapes learning's very essence, emphasizing adaptability, resilience, and perennial curiosity.

The Changing Role of Teachers: Reflecting upon the metamorphosis of educators as they transition from traditional roles to guides, facilitators, and lifelong learners in this tech-forward era.

Holistic Wellness in AI-powered Learning Environments: An exploration of the human element, ensuring emotional, psychological, and moral well-being in a world increasingly mediated by screens and algorithms.

The Transformation of Learning Spaces with AI: A conceptual tour of the architectural marvels of the future—spaces crafted for a generation seamlessly integrating the tangible and the virtual.

Moral and Ethical Dimensions of AI in Education: A conversation on the responsibilities, challenges, and ethical considerations of introducing AI-enhanced experiences into our learning ecosystems.

By this chapter's end, as we intertwine insights, research, and visions of what lies ahead, we'll ponder the essence of education in this rapidly evolving landscape. It's a tale of embracing unparalleled technological potential while staying anchored in the enduring values and principles central to authentic education. Join us on this captivating odyssey.

The Nature of Learning in AI-powered Environments

In the nascent world of AI-driven education, the very essence of learning undergoes a seismic shift. As AI tailors educational content to individual needs, it provides an unparalleled opportunity for personalized learning experiences. Yet, in this context, the focus of education broadens from mere knowledge acquisition to the development of essential skills and competencies.

Historically, students were the passive recipients of knowledge. Today, they're expected to navigate an increasingly intricate, volatile, and uncertain world. Therefore, education in the age of AI is not just about teaching; it's about equipping learners with a reliable compass—both cognitive and ethical—to steer through these complexities.

Emphasis should be on fostering a mindset that doesn't just passively accept established wisdom. Instead, learners must be encouraged to question, critique, and extend it. This is particularly evident in the realm of science education. Instead of treating science as a set of immutable facts, we must view it as a dynamic field, where understanding the methodologies and the quest for the right questions is as vital as knowing the answers.

Additionally, given the inundation of information in the digital age, it's paramount to cultivate discernment. Probabilistic thinking, which underpins many of today's challenges—from climate change to pandemic evolution—must be ingrained in the learners. They should be adept at finding signals amidst the noise, understanding relative likelihoods, and visualizing alternative futures.

In essence, in AI-powered environments, learning should be envisioned as a dynamic, evolving journey—a process of continuous exploration, questioning, and adaptation, fueled by both technological advancements and timeless human curiosity.

The Changing Role of Teachers

In the backdrop of the digital revolution and AI's inroads into education, the traditional role of the teacher is undergoing a profound transformation. No longer are educators merely the primary source of knowledge; instead, they are becoming pivotal guides and mentors in a landscape awash with information.

FIGURE 4. *No longer are educators merely the primary source of knowledge; instead, they are becoming pivotal guides and mentors.*

As the concept of literacy evolves, the role of the teacher becomes even more nuanced. In the past, literacy was about reading, writing, and arithmetic, with teachers often pointing students to set, trusted resources like encyclopedias. Today's digitally savvy students, however, face a deluge of information. Every online query presents a myriad of answers, without clear delineations between fact and fiction. In such an environment, educators become critical beacons, helping learners discern quality content, triangulate viewpoints, and comprehend nuances.

Moreover, there's a pressing need for educators to shift from a prescriptive teaching style to one that nurtures critical thinking. Rather than disseminating information, they should facilitate students in critically evaluating, contextualizing, and even creating knowledge. This transition places a heightened emphasis on educators not just as knowledge providers, but as value instillers and role models.

Furthermore, the age of AI demands that educators be adept at integrating technology into pedagogy, making learning more interactive, immersive, and tailored. Their role expands to being curators of educational experiences, ensuring that AI tools are used judiciously to foster meaningful, holistic learning experiences.

In this new paradigm, educators are tasked with a dual responsibility: to stay abreast with rapid technological advancements and to nurture learners who are not just knowledgeable but are also ethical, empathetic, and discerning global citizens.

FIGURE 5. *There's a pressing need for educators to shift from a prescriptive teaching style to one that nurtures critical thinking.*

Holistic Wellness in AI-powered Learning Environments

As literacy evolves, the role of the teacher becomes even more nuanced. In the past, literacy was about reading, writing, and arithmetic, with teachers often pointing students to set, trusted resources like encyclopedias. Today's digitally savvy students, however, face a deluge of information. Every online query presents a myriad of answers, without clear

- **Physical:** Nourishing a healthy body through exercise, nutrition, sleep, etc.
- **Mental:** Engaging the world through learning, problem-solving, creativity, etc.
- **Emotional:** Being aware of, accepting and expressing our feelings, and understanding the feelings of others.
- **Spiritual:** Searching for meaning and higher purpose in human existence.
- **Social:** Connecting and engaging with others and our communities in meaningful ways.
- **Environmental:** Fostering positive interrelationships between planetary health and human actions, choices and wellbeing.

FIGURE 6. *Wellness is Multidimensional.*

During the rapid digital evolution, with AI at its forefront, there's a rising need to approach learning from a holistic perspective. As the very nature of knowledge acquisition changes, there is a growing realization that education is not just about academic achievements; it's about nurturing the whole individual.

This idea of holistic wellness in learning environments brings forth the importance of synthesizing various fields of knowledge. There is compelling evidence on how modern societies thrive by drawing connections between seemingly unrelated ideas. It's not enough for learners to be experts in silos; they need to be adept at integrating diverse knowledge streams, bringing together the arts and sciences, the theoretical and the practical. This approach to learning fosters cognitive flexibility, creativity, and innovative problem-solving.

Furthermore, an essential component of holistic wellness is the cultivation of 'bridging social capital'. In an increasingly interconnected world, learners must be equipped to navigate diverse cultures, perspectives, and ideologies. They should be capable of understanding, empathizing, and collaborating with peers from varied backgrounds and traditions. This goes beyond mere tolerance to a genuine appreciation of diversity, helping students to develop a more expansive worldview.

The emphasis on social and emotional learning (SEL) resonates strongly within this framework. Beyond cognitive skills, students' emotional and social well-being becomes paramount. How they feel, how they relate to their peers, how they cope with challenges, and how they express themselves are all vital facets of their education. The environment should thus be designed to be responsive to these emotional and social needs, ensuring a nurturing, inclusive, and supportive ambiance.

FIGURE 7. In an increasingly interconnected world, learners must be equipped to navigate diverse cultures, perspectives, and ideologies.

FIGURE 8. Beyond cognitive skills, students' emotional and social well-being becomes paramount.

The AI-powered learning environment of the future can't just be technologically advanced; it must be human-centric, aiming to nurture well-rounded individuals who thrive not just academically but also emotionally and socially.

The Transformation of Learning Spaces with AI

With the infusion of AI, Virtual Reality (VR), Augmented Reality (AR), and other immersive technologies, we are witnessing a profound metamorphosis in traditional educational environments. These advancements are altering not only the modes of teaching and learning but also the physical realms within which they occur.

For architects, this evolution towards an immersive and digitally augmented pedagogy heralds a fresh chapter in educational space design. Here's a comprehensive understanding of what this transformation implies:

1. **Flexible Layouts:** To accommodate the dynamic nature of VR, AR, and AI experiences, we require spaces that prioritize adaptability over fixed configurations. This means having open areas that can be swiftly reconfigured to meet various technological requirements.

2. **Infrastructure Considerations:** Modern learning demands robust infrastructure—rapid connectivity, provisions for equipment storage, and specialized zones to mount or display AR/VR tools (Figure 7).

3. **Sound and Light Control:** Given the immersive nature of these technologies, the ability to manipulate ambient light and minimize external noise becomes essential. Adjustable lighting and soundproofing elements will be crucial in these spaces.

FIGURE 9. To accommodate the dynamic nature of VR, AR, and AI experiences, we require spaces that prioritize adaptability over fixed configurations. In fact, since students' primary experiences will be virtual, the room itself should be basic and with sufficient room for movement. This is one of the few rooms in school where windows are not desirable. See it as a portal to a big, bright and open world with endless possibilities and limitless experiences.

4. **Safety and Accessibility:** Design considerations must cater to unobstructed movement, inclusivity, and overall safety, especially when students engage deeply with technology.

5. **Multi-sensory Spaces:** Future spaces will appeal to more than just the auditory and visual senses. They could integrate tactile feedback tools or even olfactory stimuli to augment learning.

6. **Collaborative Areas:** AR, VR, and AI often promote collective learning. Thus, areas that encourage group interaction, discussion, and shared technological exploration will be integral.

7. **Outdoor Integration:** AR's potential to turn external spaces into vibrant learning arenas implies that areas like courtyards and gardens might play a significant role in the learning journey.

FIGURE 10. Areas that encourage group interaction, discussion, and shared technological exploration will be key to the design of spaces that will cater to the upcoming AI revolution.

FIGURE 11 and FIGURE 12. Take every opportunity to turn external spaces into vibrant learning arenas.

8. **Sustainability:** The energy consumption of these technologies accentuates the need for sustainable architectural solutions, such as solar integration, efficient cooling, or natural ventilation.

9. **Aesthetic and Inspirational Design:** Spaces should transcend functionality, igniting curiosity and mirroring the groundbreaking spirit of the tools within.

10. **Futureproofing:** It's pivotal to anticipate subsequent technological evolutions. Designing with adaptability in mind ensures spaces remain relevant and won't require extensive overhauls with each technological wave.

11. **Attention Restoration Spaces:** Given the anticipated increase in screen time due to immersive tech, there's a pronounced need for spaces designed to counteract digital fatigue. Infusing areas with natural elements, offering quiet zones, and facilitating direct interaction with nature can significantly bolster cognitive recovery, well-being, and learning outcomes.

FIGURE 13. To counteract digital fatigue. Infusing areas with natural elements, offering quiet zones, and facilitating direct interaction with nature can significantly bolster cognitive recovery.

Table 1. Critical considerations for architects in designing AI-augmented educational spaces.

Element	Description
Flexible Layouts	Open, adaptable spaces for varied tech setups.
Infrastructure	High-speed connectivity, equipment storage, specialized mounting/display areas.
Sound & Light Control	Adjustable ambient controls to optimize immersion.
Safety & Accessibility	Unobstructed movement and inclusivity.
Multi-sensory Spaces	Catering to auditory, visual, tactile, and olfactory senses.
Collaborative Areas	Spaces for group interaction and tech exploration.
Outdoor Integration	Utilizing external spaces for AR-based learning.
Sustainability	Energy-efficient solutions for tech demands.
Aesthetic Design	Inspiring curiosity and technological immersion.
Futureproofing	Spaces adaptable to technological progress.
Attention Restoration	Natural, quiet zones to counteract screen fatigue.

Moral and Ethical Dimensions of AI in Education

The integration of AI into education brings forth a series of moral and ethical challenges that must be carefully addressed:

1. **Data Privacy and Security:** Ensuring that personal student data is handled responsibly, with robust protection against breaches and unauthorized access.
2. **Algorithmic Biases:** Taking measures to prevent AI systems from perpetuating or exacerbating societal biases, leading to unequal opportunities.
3. **Over-reliance on Technology:** Striking a balance between leveraging AI's capabilities and maintaining human interaction and critical thinking.
4. **Echo Chamber Effect:** Avoiding personalized learning tools that limit exposure to diverse perspectives, potentially inhibiting critical thinking.
5. **Equal Access and Equity:** Ensuring that all students have equitable access to AI tools, irrespective of socio-economic backgrounds, to avoid widening educational disparities.
6. **Transparency and Accountability:** Maintaining clarity in how AI tools operate, especially when evaluating student performance, and preserving human oversight.
7. **Teacher and Student Autonomy:** Preserving the human elements of intuition and choice, so that teachers and students don't feel entirely dictated by algorithms.

Conclusion: Navigating the Confluence of AI and Authentic Education

In the advent of the AI era, education stands at a transformative juncture. As we've journeyed through the evolving nature of learning in AI-powered environments, the metamorphosis of teaching roles, the revitalization of learning spaces, and the crucial ethical considerations, it's evident that AI isn't merely a tool—it's a force reshaping the very edifice of educational experiences.

The essence of education, however, remains unaltered in its core mission: to nurture informed, empathetic, and versatile individuals ready to face the world's complexities. In this light, the tools, methodologies, or spaces we utilize are conduits to achieve this goal. As AI further integrates into our educational systems, it offers both profound opportunities and intricate challenges.

Harnessing AI's potential requires a holistic approach, transcending the confines of traditional learning and acknowledging the emotional, social, and ethical dimensions of human development. As outlined in our vision of future learning spaces, the architectural realm will need to adapt, integrating flexibility, sustainability, and attention-restorative spaces to cater to the demands of an AI-augmented pedagogy.

Yet, in this confluence of technology and education, educators, policymakers, and architects alike must remain vigilant. The moral and ethical implications of AI in education, while multifaceted, underscore the imperative of placing students' well-being at the forefront. It serves as a reminder that while AI can process, predict, and even teach, it's the human touch, the authentic experiences, and the quest for meaning that truly define the educational journey.

As we look ahead, the pillars of SEL, Values, Meaning, and Experience stand tall, guiding our path. In championing these values, we ensure that education, even in the most technologically advanced era, remains an endeavour rooted in humanity, growth, and authentic discovery.

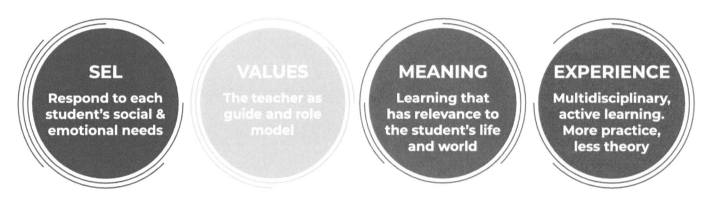

FIGURE 14. In the world of AI, it is even more important to focus on the human dimensions of growth.

References

Advance HE. (2019). Future learning spaces: Space technology and pedagogy. https://www.advance-he.ac.uk/knowledge-hub/future-learning-spaces-space-technology-and-pedagogy

Carvalho, L., Martinez-Maldonado, R., Tsai, Y.-S., Markauskaite, L., & De Laat, M. (2022). How can we design for learning in an AI world? Computers and Education: Artificial Intelligence, 3, 100053. https://doi.org/10.1016/j.caeai.2022.100053

Hamilton, A., Wiliam, D., & Hattie, J. (2023). The Future of AI in Education: 13 things we can do to minimize the damage. Cognition Learning Group, University College London, University of Melbourne. Working Paper. Retrieved from https://osf.io/372vr/download

Leibovitz, L. (2023). Future Classroom: AI and Beyond. Visual Class. https://en.visual-class.com/blog-visual-digital-literacy

Luckin, R. (2018). Machine Learning and Human Intelligence: The Future of Education for the 21st Century. UCL IOE Press. ISBN: 978-1-7827-7251-4.

Pea, R. D., & Lu, Y. (Eds.). (2021). AI in learning: Designing the future. Springer. https://doi.org/10.1007/978-3-031-09687-7

Zhao, Y. (2018). What Works May Hurt--Side Effects in Education. Teachers College Press. ISBN: 978-0-8077-5905-9.

Dr. Parul Minhas
Prakash Nair, AIA
Louis Sirota, AIA

NEUROARCHITECTURE

Health, Happiness & Learning

> *"The field of neuroarchitecture is about understanding the fundamental ways that our environment shapes our brain, behavior, and experience, and using that knowledge to design spaces that promote health and happiness."*
> *- Veronica Galvan*

Introduction

Neuroarchitecture is the study of how the built environment impacts the nervous system and its interconnection with other bodily systems, ultimately affecting our overall well-being, including cognitive and emotional health. With the growing awareness about the impact of the built environment on our mental and physical health, there has been a rising need to consider neuroarchitecture while designing schools. Designing learning spaces based on the insights from neuroscience research for architecture can have a significant positive impact on students' academic performance, emotional well-being, and physical health. Neuroarchitecture is about designing schools that are inclusive, stimulating, and supportive of learning in ways that promote a sense of purpose, belonging, and well-being, leading to a more positive and productive educational experience for students.

What is Psychoneuroimmunology and Why is it Important?

Psychoneuroimmunology is an interdisciplinary field of study that explores the relationship between the brain, behavior, and the immune system. It examines the effects of psychological and social factors on immune function and the interactions between the nervous system and the immune system. The goal of psychoneuroimmunology is to better understand the mechanisms by which stress, emotions, and other psychological factors can affect physical health and contribute to the development and progression of diseases.

In the context of school design, considering psychoneuroimmunology can help create a physical environment that promotes positive emotional health and improves the response to stressors that affect children, ultimately leading to enhanced immunity. This requires an understanding of the impact of environmental factors on the neuroendocrine system and the immune system, as well as the interplay between psychological and physical health *(Fig.15)*. By creating spaces in the school environment that enhance emotional well-being, such as spaces for stress reduction and relaxation, schools can help to improve the neuroendocrine balance in students, which in turn promotes a stronger immune system. A comprehensive approach to school design that considers the psychological, neurological, and immunological aspects of students can ultimately lead to a healthier and more resilient student population.

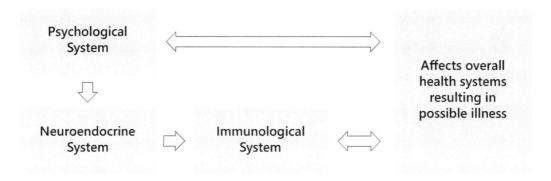

FIGURE 15. Understanding the interrelationship.

Neuroendocrine Balance And Childhood Stress

Designing schools that promote a healthy neuroendocrine balance and enhance emotional health can help combat the negative effects of stress on children. Stress can impact the physical structure of the brain, decrease immune function and lead to changes in hormone balance. It can also cause neuropeptides to be released, which can reduce the effectiveness of immune cells and damage long-term health. Aaron Antonovsky, a medical sociologist, suggested that good health starts with our ability to cope with stressors, and children need a well-functioning stress response system for healthy development. Childhood stress can lead to negative long-term health consequences, highlighting the need for investing in prevention strategies. Stress in children, when unchecked, can manifest itself in various ways as *(Table 2)*.

Table 2. Signs of Emotional Distress in Children.

THE EFFECTS OF STRESS IN CHILDREN	
Physical & Emotional symptoms	**Cognitive Symptoms**
• Irregular bowel movements • Involuntary twitching or shaking • Getting sick more often than normal • Headaches • Nausea • Muscle aches • Trouble sleeping • Heartburn or indigestion • Fatigue • Flushed skin • Clenched teeth • Unusual changes in weight • Less than normal patience • Feelings of sadness and/or depression • Feelings of being overwhelmed • Restlessness • Reduced or eliminated desire for activities once enjoyed or regularly done • Irritability • Sense of isolation • More frequent or extremely pessimistic attitude	• Impaired concentration • Trouble with remembering things, such as homework assignments or deadlines • Chronic worrying • Anxious thoughts or feelings • Reduced or impaired judgment • Impaired speech (mumbling or stuttering) • Repetitive or unwanted thoughts • Behavioral Symptoms • Change in eating habits • Change in sleeping habits • New or increased use of drugs, tobacco, or drugs • Nail biting • Abnormal failure or delay to complete everyday responsibilities • Significant change in school or work performance • Unusual desire for social isolation • Frequent lying • Trouble getting along with peers, such as classmates or teachers

RESTORING NEUROENDOCRINE BALANCE

Providing a school environment that can help restore neuroendocrine balance is essential to reducing stress in students. The right environmental stimuli can trigger desired neural responses and the secretion of necessary neurochemicals. Cortisol is a stress hormone that regulates the body's response to stress, but chronic stress can elevate cortisol levels, leading to negative physical and mental health effects. Unfortunately, over the last few decades, education has become a major source of stress for school-going children, leading to anxiety, learning disabilities, and behavioral issues. Common stressors in school include pressure to excel academically, bullying, peer pressure, social relationships, environmental stress, safety issues, poor self-esteem, and personal life disruptions.

Positive experiences in school can enhance the release of happy hormones, including serotonin, oxytocin, dopamine, and endorphins, which can alleviate stress and anxiety, boost mood, and enhance overall well-being. Serotonin and oxytocin can be increased through trust-building, social interaction, and a positive atmosphere, while endorphins are increased through laughter, music, exercise, and mindfulness practices. Dopamine can be increased by promoting learning, motivation, and a positive atmosphere. It is clear that creating a positive and supportive school environment can enhance children's sense of self-worth and belonging, leading to improved neurochemical balance and enhanced health.

Figure 16 illustrates the impact of a well-designed school environment that has the power to initiate a cycle of positive experiences. These experiences trigger the release of happy neurochemicals, such as serotonin, oxytocin, dopamine, and endorphins, leading to positive emotions and a state of balance that helps children reach their full potential. A supportive environment that promotes positive experiences contributes to children's emotional well-being and overall health, creating a positive feedback loop where children are more motivated to learn and engage in activities.

Thoughtful design interventions in schools play a pivotal role in enhancing positive experiences for students. By creating spaces that foster a sense of belonging, agency, purpose, and balance, schools actively contribute to students' overall well-being *(Figure 17)*. This initiates a cyclical process where engaging in a supportive learning environment triggers the release of happy neurochemicals, leading to improved mental and physical health.

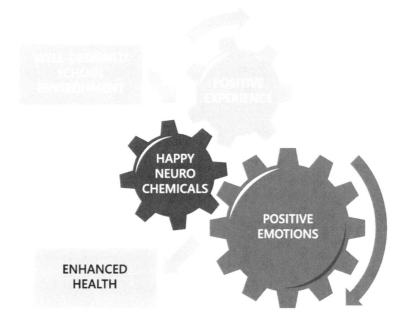

FIGURE 16. The Positive Impact of Well-Designed School Environments on Children's Health.

2 CHAPTER NEUROARCHITECTURE

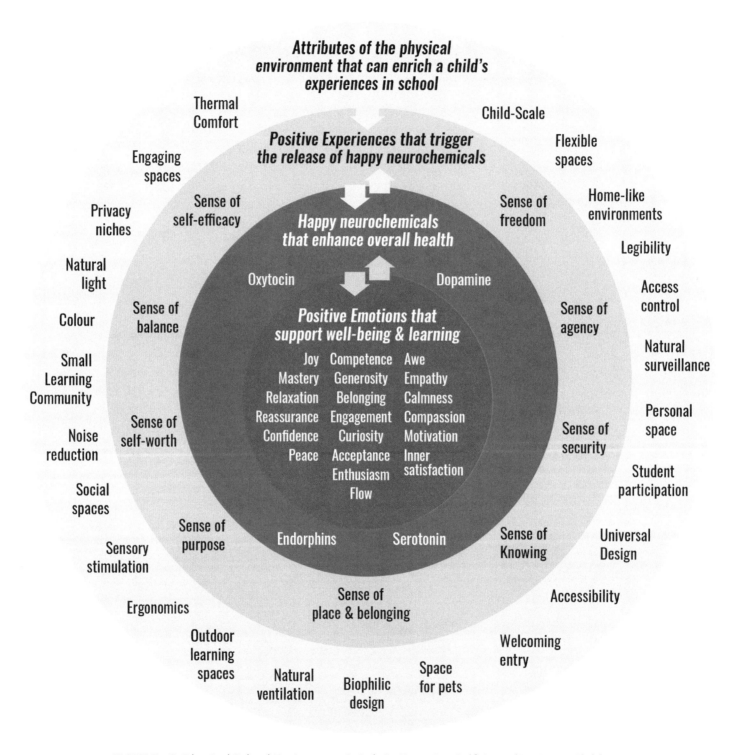

FIGURE 16. Physical School Environment's Role in Fostering Self-Actualization in Children.

NEUROARCHITECTURE INTERVENTIONS:
Enhancing Positive Experiences in School Environments

The subsequent sections will explore design interventions that can effectively increase positive experiences, creating a virtuous cycle of well-being for school children.

1. SENSE OF SELF-WORTH: A healthy sense of self-worth is an individual's perception of their value and importance. It affects their beliefs, emotions, and overall well-being. Creating an environment where students feel valued and supported can foster a positive sense of self-worth. This can trigger the release of happy hormones such as dopamine, serotonin, and endorphins, that can improve mood, reduce stress, and enhance cognitive function. A positive sense of self-worth can also lead to greater resilience and the ability to cope with challenges. School spaces can enhance a child's sense of self-worth in the following ways:

1a) Respect for scale & development needs:
Designing spaces and furnishings that are appropriate for the developmental needs of children can promote their autonomy and confidence. The scale and proportion of buildings and spaces can impact the physical and psychological comfort of occupants. To help children feel valued and engaged, spaces and furnishings should be proportionate to their size, giving them a sense of ownership and belonging. Keeping school or class sizes small can also make children feel that their efforts are significant. Weinstein (1987) recommends scaling spaces and making material storage accessible to children to enhance their self-esteem and ownership of the school environment. Here are a few specific design strategies to promote students' sense of self-worth:
- Vary ceiling heights based on the intended use of the space.
- Design spaces and specify furniture and fixtures that are proportionate to a child's scale.
- Ensure that equipment, such as whiteboards, is located at a height that children can access.
- Make material storage accessible to children.
- Install door handles, switches, and other features at a child's height.
- Offer a variety of space sizes to accommodate different needs and activities.

1b) Flexible learning spaces: Flexible learning spaces are essential to support the diverse needs of students. The physical design of schools should support various learning modalities, such as independent study, peer tutoring, and project-based learning, to name a few. These environments should also accommodate multiple intelligences in children and respect neurodiversity. Here are some ideas for creating flexible learning spaces:
- Create flexible spaces that can be easily resized or reshaped by rearranging furniture.
- Provide sufficient variety in spatial layout to support multiple modalities of learning.
- Incorporate movable partitions that can be easily adjusted.
- Use adjustable furniture that can accommodate both technology use and traditional writing/drawing.
- Install curtains or blinds to enable the use of projectors, smart boards, etc.

FIGURE 17. Design spaces and fixtures that are proportionate to a child's scale.

FIGURE 18. Create spatial layouts that support multiple modalities of learning.

1c) Ownership: Encouraging ownership and territoriality in classroom design can foster a sense of responsibility, autonomy, and belonging in students. Personalized spaces have been found to aid in memory and information retention, and classrooms that display students' work promote greater participation in the learning process. Providing each student with a "home base" or workspace can further encourage ownership and autonomy. Students also benefit from having places to display personal items, which can increase their sense of belonging in the school environment. The following ideas will improve student ownership of their learning environment:
- Personalized workspace with lockers for each student.
- Spaces designated for personal artifacts.
- Adjustable and appropriately sized desks and chairs for varying ages and sizes.
- Distinctive design elements to differentiate areas from each other.
- Display spaces for student works and projects.

1d) Stimulating playgrounds: Play allows children to develop self-reliance, self-esteem, creative and social skills, and risk-taking abilities. Risky play is necessary for children to learn how to cope with risky situations and develop risk competence. However, concerns about safety have led to a restriction on risky-play opportunities, which can hinder normal child development and lead to fear, discomfort, and dislike of the environment. Overprotection of children can have negative impacts on their health and ability to cope with unpredictability. Therefore, it is important to provide age-appropriate risky play opportunities to help children develop and thrive. Here are some ideas to create stimulating playgrounds:
- Encourage tree climbing and offer innovative play equipment with movable parts.
- Provide ample space for age-appropriate activities such as running and jumping.
- Ensure the presence of safety nets and other safety measures to prevent injuries.
- Place the school infirmary near the playground area for immediate medical assistance if necessary.
- Create playgrounds featuring a range of levels, heights, and physical challenges to encourage skill-building and risk assessment.
- Integrate natural elements, such as rocks, logs, and uneven terrain, to promote balancing, jumping, and navigating skills.
- Supply loose parts and open-ended materials, like sand, wood, ropes, or tires, to inspire creativity, problem-solving, and exploration.
- Implement a balance of supervision and freedom, allowing children to explore their environment while ensuring their safety.
- Provide age-appropriate challenges that cater to different age groups and skill levels, enabling progressive growth and development.

2. SENSE OF SECURITY: When children feel safe and protected in school, they experience positive emotions such as happiness, contentment, and well-being. This is due to reduced stress levels and the release of happy hormones such as dopamine and endorphins. A sense of security can also boost children's confidence and encourage them to take manageable risks and explore new opportunities, leading to greater engagement in their learning process. It also helps them develop healthy relationships with peers and teachers, fostering empathy, social skills, and a sense of belonging, which can further reduce stress levels and promote positive emotions. School spaces can enhance a child's sense of security in the following ways:

2a) Access control: Access control measures such as secure entry systems and visitor management systems should be incorporated into school design. Adequate lighting and visibility in and around school buildings can also help deter potential threats. A list of ideas to improve access control follows:
- Designated spaces near the entrance for safe community involvement.
- Strategically placed windows to provide clear sightlines of the entrance and outside activities.
- Effective use of signage, pavement treatments, and landscaping to direct visitors and delineate accessible areas.
- Clearly defined limits to control access and maintain a secure environment.

2b) Natural surveillance: Natural surveillance refers to the design of physical spaces in a way that allows people to see and monitor their surroundings. In the context of schools, natural surveillance can promote a sense of security in children by creating an environment that is both physically and psychologically safe. When children feel that they are being watched and protected by adults, they are more likely to feel secure and less vulnerable to threats or violence. Here are some ideas to promote natural surveillance in school:
- Install glass partitions or walls in strategic locations to increase visibility and natural light.
- Avoid creating hiding spaces in the ground, parking areas, or other places around the school through landscaping or fencing.
- Use windows and glazed doors to enhance natural surveillance of entrances, pathways, and other areas.
- Avoid unattractive barriers such as barbed wire on the school grounds. These can give a negative impression and may not be effective in preventing unauthorized access.
- If possible, attach toilets to classrooms or provide auditory connections with adjoining areas. This will prevent isolated and potentially dangerous situations.
- Install a security system with unimposing alarms, lights, and locks to provide elevated levels of security.
- Create physical or symbolic barriers along the property boundary that are attractive and welcoming.
- These can help prevent unauthorized access and create a positive impression for children and visitors.

FIGURE 19. Provide display spaces for student works and projects.

FIGURE 20. *Incorporate natural elements, such as rocks, logs, and uneven terrain that promote balancing, jumping, and navigating skills.*

FIGURE 21. *Strategically placed windows provide clear sightlines of the entrance and outside activities.*

FIGURE 22. *Glass partitions in strategic locations facilitate natural surveillance.*

3. SENSE OF FREEDOM: Providing children with a sense of freedom has numerous positive impacts on their emotional and social development. Children who have greater freedom feel less stressed and more confident, which enhances their interest and enjoyment in learning. This sense of freedom also allows children to feel respected and treated with kindness, which fosters empathy and reduces the need for disciplinary action. Additionally, a greater sense of freedom leads to positive emotions and the release of happy hormones such as dopamine and endorphins, which help reduce stress levels and promote overall well-being. Ultimately, this can help children develop social skills, a sense of community involvement, and even leadership qualities. School spaces can enhance a child's sense of freedom in the following ways:

3a) Home-like environment: Home-like school environments offer children a sense of freedom in a variety of ways. First, a warm and welcoming environment can encourage children to express themselves and take risks in their learning. Second, flexible learning spaces allow children to work in ways that best suit their individual learning style, promoting independence. Third, children are given choices and decision-making opportunities that foster autonomy. Finally, a sense of belonging and community helps children feel secure to take manageable risks and be themselves. In summary, home-like school environments promote freedom through a welcoming environment, flexibility, autonomy, and comfort. Here are some ideas to promote home-like environments:

- Choose comfortable furniture like a couch or large armchair that is inviting for children to relax and learn in.
- Incorporate nontoxic plants in the space to bring a sense of calm and freshness.
- Use natural or soft lighting from windows or lamps to create a warm and welcoming atmosphere.
- Add throw pillows and cushions to the seating area for extra comfort and coziness.
- Include decorative touches like area rugs or repurposed furniture to give the space character and personality.
- Display family photos from children and staff to create a sense of belonging and community.
- Use inexpensive frames to showcase children's artwork on the walls and encourage creativity.
- Paint the walls in neutral or pastel colors to create a calm and soothing environment that is not visually overwhelming.

3b) Density & Personal space: Considerations like density and personal space in learning environments can promote a sense of freedom in school children. When children have their own personal space and low density, they can feel a sense of autonomy, safety, and respect for others' personal boundaries. They can also feel more relaxed and focused, leading to a greater willingness to engage in their learning process.

Ultimately, promoting a culture of empathy and kindness can foster a sense of freedom and belonging for all students, contributing to a positive and inclusive learning environment. Ensure that children have ample space to move around, so that they feel less confined and more comfortable. Make sure that there is at least 7-10 sq.m. gross area provision per child to avoid overcrowding.

- Divide children into smaller groups or cohorts, which allows for more individual attention and a sense of belonging within the group.
- Keep classrooms and walls uncluttered to provide a clear and calm environment. Do not make the walls too busy or visually overwhelming.
- Limit the number of students per class to between 17-25, allowing for a more manageable and personalized learning experience.
- Consider splitting the school into smaller learning communities, which promotes a sense of identity and belonging for students within their specific community.
- Increase the area available for children by getting rid of hallways and incorporating that space into the learning community.

4. SENSE OF KNOWING: A sense of knowing refers to the ability to comprehend something instinctively, without conscious reasoning. This can instil confidence in children, empowering them to navigate their surroundings independently and reduce stress levels by enabling them to anticipate their environment. Cultivating this sense can also lead to greater emotional intelligence, self-reliance, and resilience. Moreover, a strong sense of knowing can reduce stress and promote positive emotions and happy hormones, such as dopamine and endorphins, which contribute to a sense of well-being. School spaces can enhance a child's sense of knowing in the following ways:

4a) Enhancing Legibility: Improving legibility in school environments through effective wayfinding strategies can enhance children's sense of knowing by creating a clear and organized learning environment. It can increase their confidence, motivation, engagement, and positive attitude towards learning while reducing distractions and improving their focus. Legibility plays a crucial role in shaping children's perception of their school environment and can impact their learning experiences and outcomes. Some ideas to enhance legibility in schools are as follows:

- Create a unique identity for each location within the school.
- Use landmarks as visual cues to help students identify different areas within the school.

FIGURE 23. Decorative touches with rugs and furniture can enhance a home-like school environment.

FIGURE 24. Ample space to move around makes children feel less confined and more comfortable.

- Ensure that paths within the school are well-structured and have clear goals to help students understand where they are going.
- Limit the number of navigational choices available to students to reduce confusion and increase clarity.
- Use sightlines to show students what's ahead and help them anticipate their next steps.
- Implement color-coded indoor pathways to assist students in keeping their orientation towards important locations within the learning environment.
- Ensure that the main building is an obvious point of reference among the school's buildings and that paths and buildings connect effectively and intuitively.
- Use various forms of visual cues to distinguish between different areas within the school by varying colors, textures, forms, and ceiling heights.

4b) Welcoming Entry & Signature Elements:

The school entrance should be inviting, secure and visible, with a covered space for drop-offs and pick-ups. A signature element reflecting local culture/history/architecture can enhance the welcoming feeling and promote a sense of place among children, helping them take pride in their school's unique identity. These elements should be visible not only at the entrance but also in the layout of spaces and activities. Here are some ideas to design a welcoming entry:

- Design the entrance to be inviting, highly visible, and well-defined with architectural features, signs, lighting, artwork, landscaping, and other local landmarks.
- Ensure that the entrance is not intimidating for children and is scaled appropriately for their size.
- Use motivational signs that send positive messages and make the school more inviting.
- Include landscaping features or small play areas that are visible from the entrance to make the space more inviting.
- Provide a covered entrance that provides shelter from bad weather and facilitates transition between the indoors and outdoors.
- Ensure a safe drop-off/pick-up area where children don't have to cross traffic and separate access for students and visitors.
- Incorporate signature elements relating to local culture and architecture into the built environment, preferably visible from the entrance to make it unique and special to the school.

FIGURE 25. *Use visual cues to distinguish between different areas within the school, such as colors, textures, forms, and ceiling heights.*

5. SENSE OF AGENCY: When individuals feel like they have control over their lives and can make meaningful decisions, they experience a sense of empowerment and accomplishment. This feeling of agency triggers the release of happy hormones, such as serotonin and dopamine, that can enhance mood and reduce stress levels. Furthermore, having agency allows individuals to better cope with difficult situations, as they are better equipped to make choices and take actions that can help them navigate challenges. This sense of control that children have can lead to increased resilience and greater satisfaction with life. School spaces can enhance a child's sense of agency in the following ways:

5a) Student Participation: Student participation in the school design and renovation process enhances their sense of agency by giving them ownership, fostering trust and collaboration, reducing vandalism and anti-social behavior, building competence, and raising learners' self-esteem. By working alongside industry practitioners, students can learn about risk assessment, waste management, materials performance, design techniques, and develop soft skills such as collaboration and problem-solving. Decorating their own spaces with their art and ideas can further enhance their sense of ownership and pride in their school environment. Here are some ways to increase student participation:

- Foster a sense of community and teamwork and encourage effective communication by involving students during the design of school renovation, refurbishment, and rebuilding projects.
- Consider and implement relevant student feedback and ideas by using surveys, focus groups, or other methods to gather input during the planning, design, and execution of capital projects.
- Keep students informed and engaged by providing regular updates and communications through newsletters, emails, and announcements.
- Utilize social media and other technology platforms to engage with students and encourage them to share their ideas and feedback.
- Provide opportunities for student leadership and mentorship by assigning leadership roles to them and pairing students with industry professionals.

5b) Sustainable architecture: Sustainable (green) architecture of schools can promote a sense of agency in children by involving them in the design and implementation of environmentally sound building projects. This empowers students to take ownership of their environment, fosters responsibility for their actions, and inspires them to make more conscious choices. By learning about sustainable practices and technologies, students can develop the knowledge and skills needed to become leaders in sustainability efforts in their community. Working together towards a common goal also promotes teamwork and leadership skills, leading to a more engaged and empowered student body and a

2 CHAPTER NEUROARCHITECTURE

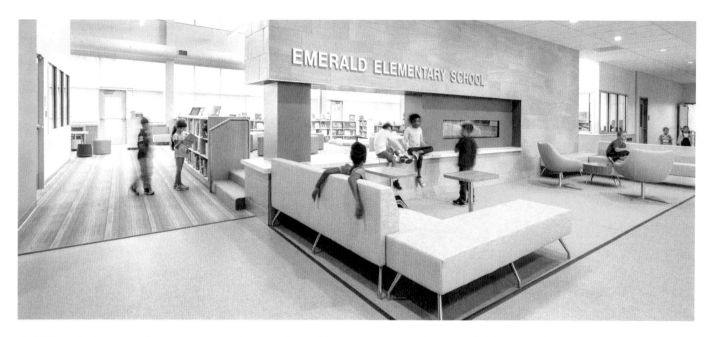

FIGURE 26. *A welcoming entry space that sends positive messages and invites children to school.*

more sustainable future for society. To promote a greater sense of agency in children, school spaces can be designed sustainably, and here are a few ideas to achieve this:

- Design spaces that allow students to learn from natural processes, such as sun orientation and wind flow patterns.
- Make energy conservation and sustainable measures visible, such as rainwater harvesting and solar panels.
- Use sustainable design features as teaching tools for project-based, experiential learning and to connect students with curricula in environmental and STEM education.
- Involve students in the design and implementation of sustainable building projects, empowering them to take ownership of their environment and fostering a sense of responsibility for their actions.

FIGURE 27. *Encouraging teamwork and fostering agency through student participation.*

FIGURE 28. Diverse, adaptable, and user-friendly furniture.

6. SENSE OF SELF-EFFICACY: Sense of self-efficacy refers to an individual's belief in their ability to achieve goals and perform tasks successfully. A strong sense of self-efficacy can help reduce stress in school children and trigger the release of happy hormones such as dopamine, serotonin, and endorphins. This leads to positive and productive learning experiences. Building a sense of self-efficacy in children involves providing them with opportunities to take ownership of their learning, setting and achieving goals, and providing positive feedback and reinforcement. When children feel empowered and confident in their abilities, they are more likely to engage in learning, take on new challenges, and experience a sense of accomplishment, leading to increased well-being. School spaces can enhance a child's sense of self-efficacy in the following ways:

6a) Universal Design Principles: Universal design principles can improve the sense of self-efficacy in school buildings by creating inclusive and accessible learning environments that support the diverse needs of all students. When students feel that their environment is accommodating and accessible to their needs, they are more likely to feel empowered and confident in their abilities, leading to a stronger sense of self-efficacy. Some ways that universal design principles can be applied in school buildings to improve self-efficacy include:
- Design clear and easy-to-follow circulation paths with tactile cues.
- Improve accessibility and hygiene of washrooms.
- Ensure appropriate width and clearance of circulation routes, and utilize easy-to-use hardware and controls.
- Consider acoustics and lighting for people with visual impairments.
- Use contrasting textures and borders to indicate changes in grade or landscape.

6b) Ergonomics: Ergonomic considerations in school environments foster self-efficacy by improving students' physical comfort and well-being. When children are free from discomfort caused by poor posture or ill-fitting furniture, they can better focus on learning, develop confidence in their abilities, and feel more empowered to take on challenges, leading to an enhanced sense of self-efficacy. Some ideas to create ergonomic learning spaces are noted below:
- Provide diverse, adaptable furniture that's user-friendly.
- Ensure age-appropriate, well-maintained furniture.
- Design IT-compatible workstations.
- Offer floor seating and reclining options in classrooms and other learning spaces.

FIGURE 29. *Designated spaces for self-reflection can be designed while allowing natural surveillance.*

7. SENSE OF BALANCE/STABILITY: Emotional stability involves the ability to remain calm, manage emotions in a healthy manner, and cope properly with stress and challenges. Emotionally stable individuals possess self-awareness, self-regulation, and self-compassion, enabling them to handle frustrations and delays, and adapt to situational demands. This stability can trigger the release of happy hormones, further enhancing well-being. Emotionally stable children effectively adjust to their surroundings, peers, and family, while understanding and expressing emotions constructively. School spaces can enhance a child's sense of balance in the following ways:

7a) Spaces for self-reflection: Self-reflection and privacy play crucial roles in promoting balance and stability in school children. Providing various spaces, including those for solitude, allows children to recharge mentally, process experiences, and achieve a sense of belonging. Opportunities for self-reflection enhance understanding and personal growth, helping students transform learning into knowledge and wisdom. Spaces for privacy and self-reflection are essential in learning space design, especially for neurodiversity, as some children may require such areas more frequently to maintain their emotional balance. Some ideas for spaces that encourage self-reflection are listed below:

- Create designated spaces for solitude, such as reading, quiet, reflection, and listening areas.
- Design classrooms and learning areas with 2-3 private spaces for controlled interactions.
- Provide inviting, supervised "cave" spaces for students to momentarily escape from busy schedules.
- Incorporate distinct breakout zones within or attached to classrooms, avoiding placement in areas with high pedestrian traffic or noise.

7b) Color for Comfort: Color influences children's interactions with their environment, impacting their sense of balance and stability. Balancing stimulation is crucial for learning environments. Brightness and warm colors can stimulate creativity, but overstimulation may result in adverse physiological and psychological reactions. Under-stimulation may cause anxiety and concentration difficulties. Children respond well to nuanced colors, especially those found in nature and human skin tones. Avoiding a neutral approach, carefully chosen colors in educational architecture, including calming colors for neurodivergent children, can enhance students' emotional stability. Some ideas for proper use of colors are listed below:

- Incorporate a variety of colors, ensuring overstimulation is avoided.
- Consider color and lighting together in the design of learning spaces.
- Add colorful displays to walls and doors.
- Apply cooler colors in physical activity areas, such as gyms and yoga halls.
- Paint privacy niches and withdrawal areas in cooler colors.
- Utilize a nature-inspired color palette.

FIGURE 30. *A variety of nature-inspired colors avoids overstimulation.*

7c) Natural Light: Natural light plays a crucial role in fostering a sense of balance and stability in school children by triggering the release of happy hormones. It offers numerous health benefits such as reducing mental fatigue, alleviating seasonal depression, and promoting relaxation. Daylight provides a dynamic light source that reduces monotony and maintains focus. Moreover, natural light contributes to the production of vitamin D, regulates internal body clocks, reduces stress and headaches, and positively impacts mood, promoting overall emotional well-being. Some ideas to improve students' exposure to natural light are as follows:
- Ensure glare-free, diffused daylight in spaces where children spend extended periods of their time.
- Install eye-level windows for views and incorporate skylights or clerestory windows to provide glare-free light throughout the space.
- Offer unrestricted views to reduce eye strain and connect indoor and outdoor environments.
- Supplement daylight with electric lighting, combining natural and artificial light sources, ideally from two sides of each room.
- Avoid direct exposure to bright light sources that can cause glare and discomfort.
- Avoid windows on East and West facades which increase glare and force teachers to close blinds.
- Allow diffused daylight from multiple directions to minimize shadows and evenly distribute light.

7d) Noise reduction: Everyday environmental noise can increase stress, resulting in increased blood pressure, and heart rate, potentially leading to lasting health issues. Younger children are particularly sensitive to noise and require quieter spaces for optimal learning. Spending time in quieter areas not only minimizes stress but also triggers the release of happy hormones, fostering a sense of balance and stability. Implementing noise reduction strategies such as those listed below can lead to improved cognitive functioning and balance restoration:
- Select a site located in a peaceful area with low noise and air pollution.
- Incorporate barriers and buffers to counter traffic noise.
- Utilize landscaping as a noise dampener.
- Avoid using hard materials that cause echoes in interior spaces.
- Use appropriate sound-absorbing materials in classrooms and common areas to minimize noise levels.

2 CHAPTER NEUROARCHITECTURE

FIGURE 31. *Ample daylight and views reduce eye strain and connect indoor and outdoor environments.*

8. SENSE OF BELONGING: A sense of belonging in schools is crucial for students' well-being and academic success. It fosters feelings of acceptance, connectedness, and being valued within the school community. A sense of belonging positively impacts academic achievement, mental health, and self-esteem. When children feel they belong, they are more engaged, motivated, and resilient. The release of happy hormones, such as oxytocin, serotonin, and endorphins, is triggered by a sense of belonging, which further contributes to their overall happiness and well-being. Ensuring a supportive and inclusive environment helps children develop into healthy, well-adjusted adults. Sense of belonging can be enhanced in school spaces through interventions like:

FIGURE 32. *Sound-absorbing ceiling, wall and free-standing panels can greatly minimize noise levels.*

8a) Spaces for peer/teacher interaction: School spaces that promote peer and teacher interaction, such as auditoriums, amphitheatres, media centres, and dining areas, can significantly enhance children's sense of belonging. These inviting and comfortable settings encourage social interaction, foster a sense of community, and facilitate collateral learning. By supporting the development of social skills and fostering enduring attitudes, these spaces help students feel connected and valued, contributing to their overall well-being and desire for lifelong learning. Some design ideas for spaces that foster peer and teacher interaction in school are noted below:
- Provide multiple safe and visible locations where groups of four or more children can interact and engage in various activities.
- Create designated areas where students can share meals with their peers and teachers, fostering a sense of community.
- Include facilities such as auditoriums or large gathering areas indoors as well gathering stairs and outdoor amphitheaters in the school design to encourage gatherings, events, and performances.

8b) Materials & Textures: Materials and textures play a crucial role in enhancing students' sense of belonging in school. By incorporating natural materials, warm colors, and diverse textures, the environment becomes more inviting and nurturing. Introducing soft, cozy elements like pillows, plants, and upholstered furniture provides a home-like atmosphere, fostering a sense of comfort and connection. The use of curvilinear forms, color gradations, and blended textures further enriches the space, making it more appealing to children. Overall, thoughtful material and texture choices contribute to supportive learning environments that promote a strong sense of belonging among students. Some ideas in this regard are noted below:
- Incorporate natural materials and showcase visible details.
- Employ multi-sensory materials to create surfaces with varying qualities such as smoothness, roughness, brightness, opacity, and transparency.
- Choose materials that evoke warmth and comfort.
- Utilize glass to establish a connection between indoor and outdoor spaces while ensuring a sense of security for students.
- Introduce soft, comforting elements like pillows, plants, and soft furnishings to create a warm, home-like atmosphere.
- Incorporate curvilinear shapes and forms whenever possible to enhance the appeal of the environment.

FIGURE 33. Informal spaces for peer/teacher interaction tend to foster a sense of belonging in students.

FIGURE 34. Curvilinear shapes and multi-sensory materials to enhance the appeal of the environment.

8c) Small Learning Community: The Small Learning Community Model enhances children's sense of belonging in school by creating an environment that promotes close peer interactions and strong bonds with teachers. These smaller communities, accommodating 80 to 150 students, prevent feelings of isolation and anonymity. The model incorporates learning studios, small group rooms, multi-purpose labs, commons spaces, and outdoor connections to support diverse learning modalities. By fostering more intimate and supportive educational settings such as those listed below, small learning communities significantly improve student engagement, performance, and overall well-being:

- Divide the school into smaller learning communities using separate blocks or levels.
- Provide each learning community with designated open spaces and indoor areas for socialization.
- Ensure transparency within each community to always facilitate passive supervision.

FIGURE 35. Small Learning Communities facilitate collaborative learning and socialization.

9. SENSE OF PLACE: "Sense of place" refers to the emotional and psychological connection that an individual has to a specific location, and it can be an important aspect of personal identity and well-being. In the context of a school, sense of place refers to the connection that students, teachers, and staff have to the physical and social environment of the school. A strong sense of place can reduce stress and anxiety in school children, as it triggers the release of happy hormones such as serotonin, dopamine, and oxytocin, leading to increased well-being and positive emotions. It can also improve academic performance and social connections, promoting a sense of belonging and enhancing personal identity. A sense of place in a school can be fostered through various means, such as:

9a) Outdoor Learning Spaces: Outdoor learning spaces in schools can help foster a sense of place by allowing students to connect with their local environment and appreciate their surroundings.

These spaces provide opportunities for collateral learning, social interaction and personalization, individual mastery, and student ownership and agency. Outdoor learning can be facilitated by using a variety of structures such as trees, awnings, tents, and green-houses. Overall, outdoor learning spaces offer a rich palette of learning modalities that inspire students to customize their learning and achieve a deeper understanding of the world around them. Some ideas for enhancing outdoor learning experiences follow:

- Provide an accessible green/open space outside each classroom.
- Include an amphitheater for outdoor plays, performances, and presentations.
- Use temporary structures and benches to facilitate outdoor classrooms.
- Organize field trips, community service, and other activities that involve new places and experiences.
- Create a variety of open spaces to allow for different types of outdoor learning.

9b) Design the school to reflect local culture and history: As examples, incorporate elements of the local architecture or landscape into the design of the school, or highlight the achievements and contributions of local figures in the school's history (refer also to section 4b).

9c) Creating opportunities for students to personalize their environment: As an example, allowing students to decorate their classroom with their own artwork, or encouraging teachers to add personal touches to their classrooms (refer also to section 3a).

FIGURE 36. Shaded outdoor spaces provide opportunities for multi-sensory learning.

FIGURE 37. *Display spaces for children's artwork can help enhance a sense of place.*

10. SENSE OF PURPOSE: A sense of purpose in school can trigger the release of happy hormones in children, promoting a sense of balance and stability. When students feel that their actions and efforts in school are meaningful and aligned with their values and goals, it leads to increased motivation, engagement, and academic achievement. This sense of purpose also creates a positive mindset, promoting mental health and well-being. When children have a sense of purpose, they feel more confident in their abilities and have a greater sense of control over their future. This can lead to the release of hormones such as dopamine and oxytocin, which contribute to feelings of happiness, fulfilment, and overall well-being. School design can augment a student's sense of purpose in following ways:

10a) Biophilic design: Biophilic design can enhance children's sense of purpose in school by providing opportunities for them to connect with nature. Exposure to natural elements can have a positive impact on children's cognitive development, emotional well-being, and physical health. By incorporating natural elements into the design of a school, children can develop a sense of wonder, curiosity, and connection to the natural world. This can help to develop a sense of purpose and stewardship in children as they learn to appreciate and care for the natural environment, leading to enhanced well-being and cognitive focus. Ideas for biophilic design in schools include:

- Incorporate natural materials such as wood, stone, and water into the design.
- Use natural colors and patterns in the interior design.
- Create spaces where natural light and fresh air flow in.
- Provide access to natural views and landscapes through windows and skylights.
- Incorporate elements such as rain gardens, green walls, and green roofs to promote biodiversity and improve air quality.
- Use furniture and other interior elements that mimic natural forms and patterns.
- Design spaces that allow for different sensory experiences, such as the sound of running water or the scent of flowers.
- Create spaces that allow for exploration and discovery, such as a nature trail and outdoor lab.
- Provide ample green and natural spaces throughout the school campus.
- Ensure that restorative spaces with soft furnishings, plants, animals, window seats, or aquariums are generously available.
- Provide opportunities for students to engage in gardening and other outdoor activities.

10b) Sensory Aesthetics: Sensory aesthetics can positively influence a student's sense of purpose in school by creating a visually pleasing and well-organized environment. The use of curvilinear forms and edges, natural elements, and a balance of visual complexity

FIGURE 38. Use natural colors and biomorphic patterns in the interior environment.

can enhance creativity and promote physical activity. Designers should also consider the different sensory needs of students and provide ample opportunities for sensory stimulation in outdoor play areas and common spaces. Adequate space for physical activities and physical education should also be included to promote physical fitness. Some ideas regarding sensory aesthetics follow:

- Create a layout that is easy to navigate and promotes a sense of flow.
- Use colors, textures, and materials that are pleasing to the eye and promote a sense of calm and order.
- Provide attractive and well-maintained landscape areas.
- Utilize highly articulated fenestrations for framing of views.
- Incorporate visually pleasing staircases and other movement pathways to encourage walking, with age-appropriate design of walking routes.
- Avoid long narrow corridors and instead use nature-connected pathways.
- Provide dedicated indoor spaces for physical activities like yoga, dance, and fitness training.
- Create sensory gardens with various activity spaces to suit the needs of children with varied temperaments.
- Provision of a sensory room or area with appropriate equipment and materials to allow students to engage in activities that promote sensory exploration and regulation.

FIGURE 39. Sensory space promote exploration and engagement.

10c) Pets in School: Pets in school can enhance children's sense of purpose by providing emotional support, empathy, and responsibility. Interacting with animals can reduce stress and anxiety while also promoting engagement and motivation for learning. Opportunities to learn about animal behavior, biology and welfare can provide a real-life context for learning, making it more meaningful for students. Pets can motivate children to take an active role in their learning and development by teaching them about animal behavior, biology, and animal welfare in a real-life context. Research has shown that pets can increase academic achievement, decrease isolation, depression, anxiety, and agitation, and teach children how to nurture, care for, and respect all life. Use these strategies to increase student exposure to animals:

- Design outdoor spaces that are safe and conducive for pets to be present, while ensuring that the safety of the children is not compromised.
- Allocate a farm area that is accessible to children and provide them with opportunities to care for and tend to the animals.
- Create a dedicated space within the school that is suitable for keeping pets, such as a pet room or an outdoor enclosure.
- Ensure that the space for pets is designed to be clean and hygienic, with appropriate ventilation and drainage.

CONCLUSION

In conclusion, the integration of neuroscience research in school design is an essential step towards cultivating healthy, happy, and successful learners. By understanding the interplay between the built environment and neurological processes, schools can actively contribute to the emotional, cognitive, and physical well-being of students. Incorporating biophilic design principles and fostering small learning communities can be the key strategies in implementing neuroarchitecture in school environments.

Biophilic design, which emphasizes the innate human connection to nature, can have a profound impact on students' well-being and cognitive function. By incorporating natural elements such as sunlight, fresh air, green spaces, and natural materials into school design, we can create environments that reduce stress, improve air quality, and enhance overall mood and cognitive performance. This approach not only creates visually appealing spaces but also supports the neurological and emotional well-being of students, enabling them to thrive academically and emotionally.

Small learning communities, characterized by close-knit groups of students and teachers, provide a supportive environment that promotes positive social interactions and a sense of belonging. These communities foster strong relationships, personalized learning, and a collaborative atmosphere, which can lead to improved academic outcomes and emotional well-being. By integrating small learning communities into the school design, we can create spaces that facilitate interpersonal connections, trust-building, and collaborative learning, ultimately enhancing the neuroendocrine balance and promoting students' overall health.

By fostering a sense of belonging, agency, purpose, and balance through thoughtful design interventions, schools can effectively promote a neuroendocrine balance, enhancing students' resilience to stress and improving overall health. A well-designed school environment with the above-mentioned considerations not only elevates academic experience but also positively impacts students' long-term health, empowering them to reach their full potential. It is crucial that architects, educators, and policymakers recognize the value of neuroarchitecture in designing schools that nurture the minds and bodies of our future generations.

References

Antonovsky, A. (1979). Health, stress, and coping. Jossey-Bass.

Antonovsky, A. (1987). Unraveling the mystery of health: How people manage stress and stay well. Jossey-Bass.

Barrett, P., & Barrett, L. (2019). Primary schools must be designed to enhance learning. In H.M. Tse et al. (Eds.), Designing buildings for the future of schooling: Contemporary visions for education (pp. 113-130). Routledge.

Cleveland Clinic. (n.d.). Serotonin. Retrieved from https://my.clevelandclinic.org/health/articles/22572-serotonin

Economic Times. (2018). Boost these hormones to succeed as a leader at work. Retrieved from https://economictimes.indiatimes.com/wealth/earn/boost-these-hormones-to-succeed-as-a-leader-at-work/articleshow/66988190.cms?utm_source=contentofinterest&utm_medium=text&utm_campaign=cppst

FOGSI. (2020). Endocrinology committee newsletter, 26. Retrieved from https://www.fogsi.org/wp-content/uploads/committee-2020-activities/vol-26-endocrinology-committee-newsletter.pdf

Gallagher, W. (1999). How places affect people: Buildings have a huge influence on our mood and performance. Why haven't architects heeded the findings of environmental behavioral science? Architectural Record, 187(2), 74.

Gardner, H. (1993). Multiple intelligences: The theory in practice. Basic Books.

Happyfeed. (n.d.). 4 brain chemicals that make you happy. Retrieved from https://www.happyfeed.co/research/4-brain-chemicals-make-you-happy

Heinrichs, M., Baumgartner, T., Kirschbaum, C., & Ehlert, U. (2003). Social support and oxytocin interact to suppress cortisol and subjective responses to psychosocial stress. Biological Psychiatry, 54(12), 1389-1398.

Hughes, H., Wills, J., & Franz, J. (2019). School spaces for student wellbeing and learning. Springer Nature Singapore.

Jain, V., & Sharma, R. (2021). The dual role of oxytocin in stress and well-being: Implications for mental health. Indian Journal of Positive Psychology, 12(2), 237-243. Retrieved from https://ijip.in/wp-content/uploads/2021/08/18.01.036.0210903.pdf

Kandhalu, P., & Kim, H. J. (2013). Effects of cortisol on the brain: A review. Berkeley Scientific Journal, 18(1). Retrieved from https://bsj.berkeley.edu/wp-content/uploads/2013/11/04-FeaturesEffects-of-Cortisol_Preethi-KandhaluKim.pdf

Kosfeld, M., Heinrichs, M., Zak, P. J., Fischbacher, U., & Fehr, E. (2005). Oxytocin increases trust in humans. Nature, 435(7042), 673-676.

Lazarus, R. S., & Folkman, S. (1984). Stress, appraisal, and coping. Springer Publishing Company.

Learning Liftoff. (n.d.). 5 school stressors that interfere with learning. Retrieved from https://www.learningliftoff.com/5-school-stressors-that-interfere-with-learning/

Minhas, P. (2022). The design of school environments to promote holistic health and wellbeing of children (Doctoral dissertation, Guru Nanak Dev University, Amritsar). Department of Architecture.

Minhas, P., & Nair, P. (2022). The design of learning environments to promote student health & well-being. Association for Learning Environments, Education Design International. ISBN 9798411710991.

Naparstek, B. (2012). Scared sick: The role of childhood trauma in adult disease. Free Press.

Nair, P. (2020). Outdoor learning: Leaving the classroom behind. White paper for Association for Learning Environments (A4LE).

Nair, P., & Fielding, R. (2007, 2020). The language of school design: Design patterns for 21st century schools. Minneapolis, Minn.: Design Share.

Olds, A. R. (1979). Designing developmentally optimal classrooms for children with special needs. In Special education and development: Perspectives on young children with special needs (pp. 91-138). University Park Press.

Oyelola, K. (2014). Wayfinding in university settings: A case study of the wayfinding design process at Carleton University. Ottawa, Ontario.

Psycom. (n.d.). Oxytocin. Retrieved from https://www.psycom.net/oxytocin

Sanoff, H. (1991). Visual research methods in design (Routledge Revivals) (1st ed.). Routledge. https://doi.org/10.4324/9781315541822

Suresh, M., Franz, J., & Smith, D. (2005). Holistic health and interior environment: Using the psychoneuroimmunological model to map person-environment research in design. In R. Goh & N.R. Ward (Eds.), Proceedings Smart Systems 2005 Postgraduate Research Conference (pp. 188-195). Queensland University of Technology, Brisbane, Australia.

Tanner, C. K. (2008). Explaining relationships among student outcomes and the school's physical environment. Journal of Advanced Academics, 19(3), 444–471.

Tanner, C.K. (2009). Effects of school design on student outcomes. Journal of Educational Administration, 47(3), 381–399.

UNM. (n.d.). Effects of human-animal interactions. Retrieved from https://www.unm.edu/~lkravitz/Article%20folder/animal_interactions.html

Walden, R. (2015). The school of the future: Conditions and processes – Contributions of architectural psychology. In The future of educational research (pp. 89-148). https://doi.org/10.1007/978-3-658-09405-8_5.

Weinstein, C.S., & Thomas, G. (1987). Spaces for children: The built environment and child development. Plenum Press.

Yeang, K., & Dilani, A. (2022). Ecological and salutogenic design for a sustainable healthy global society. Cambridge Scholars Publishing.

Zhang, Y., Research, S., & Barrett, P. (2009). Optimal learning spaces design implications for primary schools. Design & Print Group, Salford, United Kingdom.

2 CHAPTER NEUROARCHITECTURE

Prakash Nair, AIA
Dr. Parul Minhas
Anna Harrison
Karin Nakano

Learning Spaces Inspired by Nature

Introduction

The rapidly developing and urbanizing world has distanced many people from nature and natural processes that used to be a central part of human activity. In fact, by 2050, it is said that 66% of the world will be urbanized,[1] further dissociating humans from nature. While this scenario is distressing for all people, it is particularly harmful to children in their developmental years.

Disconnecting with nature comes at a heavy price for humanity. Studies have revealed that the application of biophilic design in hospitals, workplaces, and schools has demonstrated many positive outcomes. These studies concluded that biophilic environments are health-promoting and restorative. They are known to combat mental fatigue with stress recovery leading to enhanced creativity, relaxation, and excitement. Moreover, other studies have revealed that biophilia not only alters human attitude and behavior but can also positively reduce the so-called "sick building syndrome," in which people suffer from health symptoms linked to the buildings in which they spend most of their time.[2] Given the number of hours that children spend indoors in outdated school facilities, they are most at risk of suffering the consequences of an environment devoid of nature and natural elements.

Although most people nowadays do not participate in primitive activities such as hunting and are not concerned about the need to be in a space that is protected from predators, these instincts are still embedded in our brains. Our brains react to certain spaces in certain ways and feeling connected to nature is shown to have a positive impact on our well-being. When people are exposed to a natural environment, they recover significantly faster from ailments compared to when they are exposed to an urban setting.[3]

Biophilic Design invites urbanized spaces to include natural elements and processes to improve people's well-

FIGURE 40. Image Source: Prakash Nair

being inside buildings and spaces. It is an approach to design that aims to connect humans and nature in our living, learning, and working places. As the world becomes increasingly urbanized, not only adults but more children will lose their connection to nature as they grow up. Despite this fact, the basic form of schools remains to have plain walls, a limited number of windows, and made with materials that do not resemble nature. It is important to recognize how meaningful and beneficial it is for schools to incorporate biophilic designs to ensure that children, no matter their backgrounds or identity, are exposed to such designs and can improve their well-being while at school.

3 CHAPTER BIOPHILIC DESIGN

What is Biophilic Design?

Biophilic Design is a design concept that aims to connect humans and the natural world within our built environments and communities. It is the practice of incorporating nature, natural materials, and concepts into human-made environments, creating a closer connection to nature and the surrounding environment. The word "biophilia" can be broken down into two ancient Greek words: "bio," which translates to nature, and "philia," which translates to love, pointing to the love of life or living systems.[4] Biophilia refers to our innate affinity as humans toward nature and natural systems, and biophilic design recognizes this affinity by connecting people to the natural world inside buildings and in our living and learning environments.

People of all ages constantly interact with the space around them. As the world continues to urbanize with technological and industrial architecture, the fundamental connection to nature that we are hardwired to thrive in can feel lost. Biophilic design enables us to create environments that harness this basic human need to be connected to nature as an important way to improve our well-being.

Terrapin Bright Green takes a neuroscientific and psychological approach to biophilic design. They offer an accessible framework for interpreting and adopting biophilic design principles into practice, known as the "14 Patterns of Biophilic Design."[5] We will discuss this framework in detail in a later section.

14 Patterns of Biophilic Design:

Nature in the Space
1. Visual Connection with Nature
2. Nonvisual Connection with Nature
3. Nonrhythmic Sensory Stimuli
4. Thermal and Airflow Variability
5. Presence of Water
6. Dynamic and Diffuse Light
7. Connection with Natural Systems

Nature Analogs

8. Biomorphic Forms and Patterns
9. Material Connection with Nature
10. Complexity and Order

Nature of the Space

11. Prospect
12. Refuge
13. Mystery
14. Risk/Peril

FIGURE 41. Vo Trong Nghia's Farming Kindergarten
Image Source: Arquitectura Viva https://arquitecturaviva.com/works/jardin-de-infancia-dongnai-3

Biophilic Design is Not a New Phenomenon

> *Throughout our evolution, we've spent 99.9 percent of our time in nature. Our physiology is still adapted to it. During everyday life, a feeling of comfort can be achieved if our rhythms are synchronized with those of the environment.*
> —Yoshifumi Miyazaki[6]

For centuries, humans have incorporated nature themes into their structures, decorative and symbolic ornamentation, and homes and public spaces. Examples include animal figures depicted in Neolithic Göbekli Tepe, the Egyptian sphinx, garden courtyards of the Alhambra in Spain, the aviary in Teotihuacan—the ancient Mexico City, and bonsai in Japanese homes.[7] The consistency of natural themes in historic structures and spaces across cultures demonstrates how important it is for humans to maintain connections with nature. In an urbanized world, adopting the approach of biophilic design allows us to create buildings and spaces that respond to such fundamental needs as humans.

Water on site, the sound of streaming water, and the sensation of water respond to our need for drinking water. A breeze and variation in air flow imply that there is fresh air for us to breathe. Having vegetables and fruits on site, and picking and growing them by hand, respond to our need for food. A safe and contained environment with a good view of the landscape was critical for our ancestors to keep an eye out for predators or animals to hunt. Although we don't live in the same world as our ancestors did, we've still inherited such survival instincts, even if we are normally not aware of them. When these basic survival needs are met, people become calmer, exhibit more prosocial behavior, and become more comfortable and happier.[8]

FIGURE 41. *Animal figure depicted in Neolithic Göbekli Tepe*
Source: Global Heritage Fund

3 CHAPTER BIOPHILIC DESIGN

Benefits of Biophilic Design

- Students in classrooms that incorporated biophilic design scored 3.3 times higher in math assessment tests compared to those who were not over a seven-month period.[9]
- Having plants in learning spaces can lead to improved performance by 10–14% and reduce the impact of ADHD.[10]
- Students in classrooms with the most daylight tested 7–18% higher than those with the least and demonstrated 20–26% faster learning rates.[11]

Nurturing Children's Biophilia

Children have an innate tendency to explore and bond with the natural world. Incorporating natural environments into learning spaces can nurture children's love of nature. Through regular contact and play in the natural world, children can develop empathy and a love of nature, which also leads to positive environmental behaviors and attitudes.[13] Children who develop biophilia—a love for nature—will be more inclined to learn about the natural world, improve environmental attitudes, and grow up becoming conscious adults who appreciate and respect nature. What is most important is to encourage a positive attitude toward nature, grow the love of nature together, and most of all, bring love and joy to the whole experience.

 Knowledge without love will not stick. But if love comes first, knowledge is sure to follow.

—White Hutchinson Leisure and Learning Group[12]

FIGURE 42. Image Source: Prakash Nair

Science Proves the Importance of Biophilic Design

Research has shown that biophilic design can reduce stress, blood pressure, and heart rates while positively impacting emotions and moods, increasing productivity, creativity, and overall well-being.[14] Neuroscience literature suggests the human brain can easily process patterns with repeating lines in collinear, curvilinear, and radial patterns found throughout nature.[15] Veins of a leaf, branches of a tree, and ocean waves are all examples of such patterns. Research hypothesizes that this ease of processing such patterns allows people to relax and focus and including such design principles in learning spaces would benefit learners as they navigate complex academic problems.[16]

Research has shown that children are more sensitive to light exposure than adults. This is because children have larger pupils and significantly greater light-induced melatonin suppression and circadian-system sensitivity to light exposures.[17] "Circadian rhythms are physical, mental, and behavioral changes that follow a 24-hour cycle," responding primarily to light and dark.[18] Studies have indicated that higher levels of average daylight exposure per day for children are associated with reduced sedentary time both on the weekdays and weekends and with increased levels of physical activity on the weekends. Access to natural light is shown to improve subjective well-being, increase levels of alertness and cognitive processing speed, and lead to better concentration.[19] Designing learning spaces with greater exposure to natural light will help to properly maintain a learners' circadian rhythms and provide a wide range of associated health and performance benefits.

Several studies have shown the positive effects of surrounding green spaces on the learner.[20] One study conducted in an elementary school in Baltimore, Maryland, showed how students found green schoolyards as places to retreat from stress, allowing them to build competence and form supportive relationships.[21] Other studies also have shown the positive relationship between nature near schools and school-wide academic performance. These findings suggest nature allows students to have mental breaks throughout their time in learning communities, contributing to improved attentional functioning.

FIGURE 42. *Veins of a leaf are one example of patterns with repeating lines in collinear, curvilinear, and radial patterns found throughout nature.*

Biophilic Design Removes Barriers to Learning

Children spend a significant amount of time in school when not at home. It has been estimated that, globally, one in seven 10–19-year-olds experience mental health conditions.[22] Yet, these conditions remain largely unrecognized and untreated. Research has shown that adolescents with mental health conditions are prone to discrimination, social exclusion, stigma, educational difficulties, risk-taking behaviors, and physical ill-health. The primarily built environments where people live, work, learn, and recreate are known to have a direct bearing on the physical and mental health of the occupants. Schools, due to the substantial amount of time that children spend there, are important places to intentionally incorporate biophilic design as a direct way to improve their health and cognitive capacity. By incorporating biophilic design, learning spaces have the potential to positively impact generations of children, promoting early academic success, and setting them on a positive trajectory in life.[23] Biophilic designs are shown to remove barriers to learning by reducing stress, improving cognitive performance, and evoking positive emotions and moods.[24]

In a study by Determan et al., students were divided into two groups: one in a learning environment with biophilic design and another in a traditional setting. The study found that students who were in a biophilic learning environment experienced a significant reduction in their stress levels compared to those who were in the control classroom, suggesting that biophilic design helped to reduce students' stress levels during the semester.[25] Only 35% of students in the biophilic environment perceived their stress levels to be high compared to 67% of students in the control group *(Figure 42)*.[26] Students in the biophilic environment favored their space more and had higher levels of involvement compared to the control group (Figure 5).[27] Furthermore, using i-Ready test scores, this study found the biophilic design group to have higher average test scores in math compared to the control group, with the biophilic design group scoring 3.3 times higher than the control group and 7.2% more students in the biophilic group testing at grade level *(Figures 43 and 44)*.[28] Qualitative data also showed that students felt "more relaxed, calm, better able to concentrate, easier to focus and have more of a purpose to learn" in the biophilic learning environment compared to other

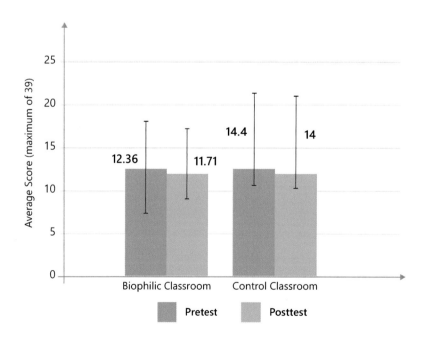

FIGURE 43. Students' opinions about their own stress level comparing the biophilic classroom (n=12) with the control classroom (n=12) at Green Street Academy, Baltimore, MD.

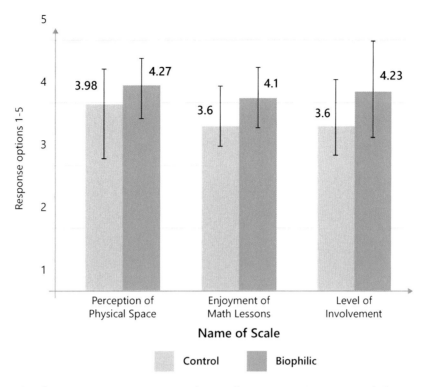

FIGURE 44. *Students' survey responses with error bars comparing a control classroom (n=17) with the purposefully designed biophilic classroom (n=16) at Green Street Academy, Baltimore, MD. Differences for all three variables are statistically significant (p<0.01).*

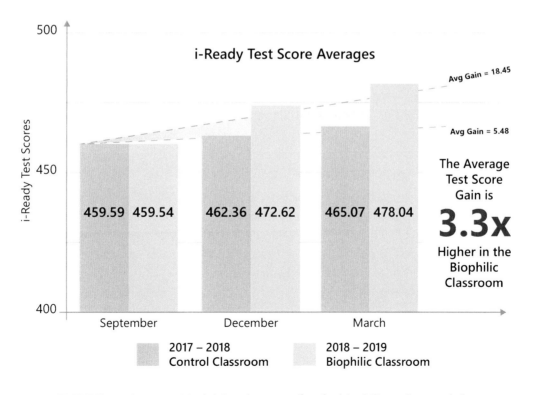

FIGURE 45. *Average Math i-Ready scores for the biophilic and control classes.*

3 CHAPTER BIOPHILIC DESIGN

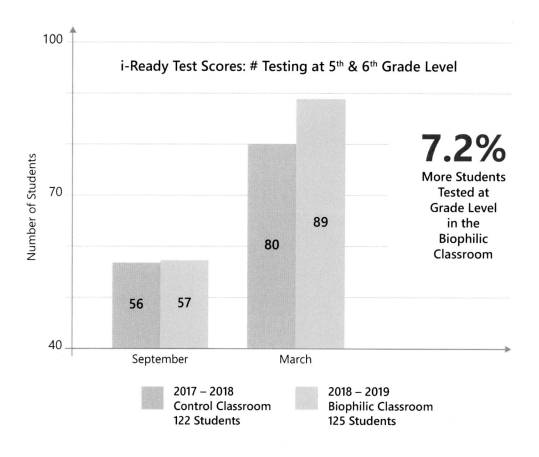

FIGURE 46. Average number of students testing at grade level.

Figures 43–46 Source: Determan, J., Akers, M. A., Albright, T., Browning, B., Martin-Dunlop, C., Archibald, P., & Caruolo, V. (2019). The Impact of Biophilic Learning Spaces on Student Success. BRIK. https://www.brikbase.org/content/impact-biophilic-learning-spaces-student-success

Obstacles to Implementing Biophilic Design

In the past decade, biophilic design has gained prominence as a critical concept. However, there may still be challenges that prevent people from implementing biophilic design:

- Lack of understanding of the significance of biophilic design.
- Perception of materials for biophilic design as being expensive.
- Competing priorities dictating function over considering design as a way of improving health and well-being.
- Not understanding the emotional, physiological, and psychological effects of physical space.
- Lack of support from coworkers and other community members.
- Concerns about maintenance of the space.
- Allergy and other health-related concerns (for gardens and other outdoor environments).
- Requiring access to expertise in design.

14 Patterns of Biophilic Design

The 14 Patterns of Biophilic Design is a framework for understanding and incorporating various biophilic design strategies into both indoor and outdoor spaces, developed by Terrapin Bright Green. It organizes biophilic design into three categories: Nature in Space, Natural Analogs, and Nature of the Space.

Nature in Space

Nature in Space addresses the direct presence of nature in a space, including plant life, animals, water, breezes, sounds, scents, and other aspects of nature.[30] It allows people to see, hear, taste, smell, and feel nature directly. "Potted plants, flowerbeds, bird feeders, butterfly gardens, water features, fountains, aquariums, courtyard gardens and green walls or vegetated roofs" are all examples of this.[31] Nature in Space includes seven design patterns:

1. Visual Connection with Nature
This pattern simply speaks to the visual presence of nature and living systems and processes in the space. Simply adding greenery to the space or having access to natural views outside the window would fit into this pattern.

2. Nonvisual Connection with Nature
Any sound, smell, taste, or touch that has a positive reference to nature and living systems belongs to this pattern. Examples would be hearing the sound of the trees bending with the wind, smelling the flowers, and touching the soil in a small pot plant. Digital simulations of nature sounds and fabric that mimic natural material textures are examples of constructed stimuli that fit into this pattern.

3. Nonrhythmic Sensory Stimuli
Nonrhythmic sensory stimuli refer to both natural and constructed stimuli that occur unexpectedly and lasts a very short time. These stimuli are "brief but welcome distraction[s]" that come across as "something special, something fresh, interesting, stimulating and energizing."[32] Moving clouds, bubbling water, movements of insects and animals, and the chirping sounds of birds are examples of naturally occurring stimuli that belong to this pattern. "Billowy fabric or screen materials that move or glisten with light or breezes, reflection of water on a surface, shadows or dappled light that change with movement or time, and natural sounds broadcasted at unpredictable intervals" are examples of such stimuli.[33]

4. Thermal and Airflow Variability
This pattern refers to the subtle changes in air temperature, airflow, humidity, and surface temperatures that imitate nature.[34] The key to this pattern is variety: Humans prefer to have access to a variety of surfaces, thermal

FIGURE 47. Average Math i-Ready scores for the biophilic and control classes.

temperatures, and airflow.[35] Things as simple as shadows and shades, controlling HVAC (heating, ventilation, and air conditioning), and opening/closing windows could be effective in making this pattern work.

5. Presence of Water
Any connection to water would tap into this pattern. This includes being able to see, hear, or touch the water in space. Examples of naturally occurring stimuli are rivers, streams, oceans, ponds, wetlands, and rainfall.[36] The presence of water can also be achieved by constructed waterfalls and streams, water walls, fountains, reflections of water, and images that include water in some form.

FIGURE 48. Image Source: Flickr Karen Mardahl
https://www.flickr.com/photos/kmardahl/14403311147/

FIGURE 49. Image Source: Flickr Donnie Ray Jones
https://www.flickr.com/photos/donnieray/16098680386/

FIGURE 50. Water play is an important element of Biophilic Design.

6. Dynamic and Diffuse Light

This pattern refers to the varying intensities of light and shadow that change over time. Daylight is dynamic as it changes color from blue in the morning to orange as it approaches sunset and then the black of night. Human bodies interact with such light and receive signals for when to be alert and when to start calming down for rest. Examples of stimuli that belong to this pattern are sunlight from different angles, direct and indirect sunlight, firelight, moonlight and starlight, bioluminescence, electric light sources, color tuning lighting, or circadian color reference that produces white light during the day and minimizes blue light at night, and dimming controls.[37]

7. Connection with Natural Systems

Natural systems refer to natural processes, such as seasonal and temporal changes that are representative of a healthy ecosystem.[38] This pattern evokes a relationship to the greater whole, promoting environmental stewardship of healthy functioning ecosystems. Climate and weather patterns, hydrology, geology, and animal behavior such as predation, feeding, foraging, mating, and habitation are examples. Pollination, growth, aging, and decomposition of insects, flowers, and plants, diurnal patterns such as shadow casting and tidal changes, night sky and cycles of the moon and other planets and stars, and seasonal patterns are also examples of systems that occur naturally.[39] These systems can be experienced through constructed stimuli such as gardening and exposure to water structure as well.

FIGURE 51. Image Source: Flickr Jennifer C.
https://www.flickr.com/photos/29638108@N06/8573417138/

Nature Analogs

Nature analogs include the use of natural materials, textures, patterns, and colors that remind us of nature. Nature Analogs include three design patterns as shown in 8, 9, 10:

8. Biomorphic Forms and Patterns

Biomorphic forms and patterns include organic shapes that use curvy lines, patterns, textures, and numerical arrangements that exist in nature. Examples include fabrics, carpet, and wallpaper designs based on the Fibonacci series or the Golden Mean, columns shaped like trees, and wallpaper patterns that represent animals and natural objects.[40]

9. Material Connection with Nature

The use of materials and elements from nature will be part of this design pattern. Humans prefer to be surrounded by the color and materials of nature. Natural materials should reflect the local ecology or geology and create their own sense of space.[41] Wood, leather, stone, fossil textures, bamboo, rattan, and dried grasses are materials that could be used in decor as well as wall construction and furniture.[42]

The use of natural colors, especially the color green is shown to have positive psychological effects on humans. Research has shown that having exposure to green before working on a task facilitates creativity and performance.[43] Certain colors are known to have certain psychological effects, as illustrated in the figure.

BLUE: clear sky or clean water - calming and relaxing

GREEN: healthy vegetation - calming and restorative

YELLOW: warmth and sunshine - happy and welcoming

RED: healthy ripe fruits - energising and exciting

FIGURE 52. *Psychological effects of four natural colors.*[44]

10. Complexity and Order

The complexity of nature has an order to it and a space with such a design pattern feels engaging and full of information with a healthy balance between being bored and overwhelmed. Research has found that "nested fractal designs with [a scaling factor of three] are more likely to achieve a level of complexity that conveys a sense of order and intrigue, and reduces stress."[45]

This pattern incorporates symmetries and fractal geometries into the space. They can be included in wallpaper, carpet, and window designs, and can be achieved through a selection of various plants and placement and exposed structure and mechanical systems.[46]

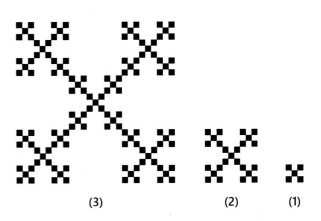

(3)　　　(2)　(1)

A square ■ with a scaling factor of 3 is more impactful than to a factor of 2.

FIGURE 53. *Image Source: Health, O., & Goode, E. (2018, June). Creating Positive Spaces Using Biophilic Design. Interface. https://www.interface.com/IN/en-IN/campaign/biophilic-design/Biophilic-Design-Campaign-en_IN*

FIGURE 54. *Shorecrest Preparatory School, St. Petersburg, Florida.*

Nature of the Space

Nature of the Space speaks to our primordial need for safety, security, excitement, and exploration and addresses the spatial arrangements in nature.[47] Humans have an "innate and learned desire to be able to see beyond our immediate surroundings" and are fascinated by the "slightly dangerous or unknown, obscured views and revelatory moments [and] sometimes even phobia-inducing properties when they include a trusted element of safety."[48] Nature of the Space includes four design patterns as shown in 11, 12, 13, 14:

FIGURE 55. *American School of Bombay.*

11. Prospect

This design pattern taps into the human's desire to see vast vistas from a safe perch for supervision and planning. Examples of this pattern include the use of "transparent materials, balconies, catwalks, staircase landings, open floor plans, elevated planes, [and] views including shade trees, bodies of water or evidence of human habitation."[49]

12. Refuge

As the name suggests, this pattern refers to spaces that allow people to withdraw from environmental conditions such as the weather, the main activity, or the flow, and feel protected. Refuge spaces can have characteristics such as a space with several sides covered—such as booth seating, reading corner, canopy beds and trees, covered walkway, porches; or a space with all sides nearly or completely covered—such as meeting rooms, private booths, and tree houses.[50] These spaces could protect people from the weather, have audio or visual privacy, or be reserved for specific purposes such as "reflection, rest, relaxation, reading, or complex cognitive tasks."[51] Such spaces have features such as adjustable or translucent shades, blinds or partitions, lowered ceiling, and varied light intensity, color, and temperature.

13. Mystery

Humans have an innate desire to explore. This pattern provides people with a sense of anticipation, denial, and reward that compels further investigation. Spaces with a mystery pattern may include small windows that only reveal parts of the space ahead, a stairway that bends over, curving edges, winding paths, and variation in brightness.

14. Risk/Peril

This pattern refers to designs that include identifiable threats coupled with a reliable safeguard. A space with this design pattern contains the possible risks of falling, getting wet and hurt, and losing control that is created by attributes such as height, gravity, water, and possible predator-prey role reversal.[52] Examples of such spaces include a climbing wall, balcony or catwalk connected to a high atrium, a railing or floor that is transparent, a path that passes under, over, or through water, and proximity to an active honeybee apiary or other animals that put humans into danger.[53]

Biophilic Design Patterns and Biological Responses

Biological responses[54] in **Table 3** illustrate how each design pattern is shown to be effective in the following three categories: stress reduction, cognitive performance, and emotion, mood, and preference.[54] Each pattern is marked with zero to three asterisks that indicate the amount of empirical data that supports the shown functions. The pattern with more asterisks implies that it is supported by more rigorous data, there is a high number of good quality peer-reviewed evidence, and has the potential for great impact.[55] No asterisk indicates that there is little research that supports the biological relationship between health and design, but there is sufficient and irresistible anecdotal information that allows us to hypothesize its importance and potential impact.[56]

FIGURE 56. A climbing wall is a perfect example of the Risk/Peril strategy. Climbing on a steep surface is an inherently risky activity that has been engineered to retain the thrill while keeping students safe.

TABLE 3. Biophilic Design Patterns and Biological Responses.

14 Patterns	*	Stress Reduction	Cognitive Performance	Emotion, Mood, and Preference
Nature In The Space				
Visual Connection with Nature	***	Lowered blood pressure and heart rate	Improved mental engagement/attentiveness	Positively impacted attitude and overall happiness
Nonvisual Connection with Nature	**	Reduced systolic blood pressure and stress hormones	Positively impacted cognitive performance	Perceived improvements in mental health and tranquility
Nonrhythmic Sensory Stimuli	**	Positively impacted heart rate, systolic blood pressure, and sympathetic nervous system activity	Observed quantified behavioral measures of attention and exploration	
Thermal and Airflow Variability	**	Positively impacted comfort, well-being, and productivity	Positively impacted concentration	Improved perception of temporal and spatial pleasure
Presence of Water	**	Reduced stress, increased feelings of tranquility, lower heart rate, and blood pressure	Improved concentration and memory restoration. Enhanced perception and psychological responsiveness	Observed performance and positive emotional responses
Dynamic and Diffuse Light	**	Positively impacted circadian system functioning. Increased visual comfort		
Connection with Natural Systems				Enhanced positive health responses; shifted perception of environment
Natural Analogs				
Biomorphic Forms and Patterns	*			Observed view preference
Material Connection with Nature			Decreased diastolic blood pressure. Improved creative performance	Improved comfort
Complexity and Order	**	Positively impacted perceptual and physiological stress responses		Observed view preference
Nature of the Space				
Prospect	***	Reduced stress	Reduced boredom, irritation, fatigue	Improved comfort and perceived safety
Refuge	***		Improved concentration, attention, and perception of safety	
Mystery	**			Induced strong pleasure response
Risk/Peril	*			Resulted in strong dopamine or pleasure responses

Source: Terrapin Bright Green. (September 12, 2014). *14 Patterns of Biophilic Design*. https://www.terrapinbrightgreen.com/reports/14-patterns/#rediscovering-the-intuitively-obvious

Link to Google Sheets: *Biophilic Design and Biological Responses* at: https://tinyurl.com/mrwkv4cj

Examples of Biophilic Designs

Vo Trong Nghia's Farming Kindergarten (Vietnam)

- This kindergarten in Vietnam has a knot-shaped roof with a vegetable garden on top and three protected courtyard playgrounds.
- The surface of the roof is covered in grass and plants to incorporate additional green into the learning space.
- By interacting with nature and growing vegetables on the roof, children learn the importance of agriculture and find connections to nature.
- One end of the slope goes down to the ground and the other rises up to the second floor of the building.

Image Source (for both images): Dezeen https://www.dezeen.com/2014/11/11/farming-kindergarten-vo-trong-nghia-architects-vietnam-vegetable-garden/

The Paul Chevallier School (France)

- The Paul Chevallier School in Lyon, France, utilizes natural materials to increase contact with nature throughout the building. Wooden cladding covers the interior and exterior walls and prompts tactile human-nature interaction, reducing stress, and providing energizing and relaxing experience.
- Both the elementary and nursery school are built in V-shapes surrounding the outside space. The green roofs, at the same time, expand the outdoor space for children to explore and interact with the outdoor environment.
- The school also encompasses a vegetable garden and faces a woodland park, which offers a view of nature from the learning communities.
- Corridors have windows that extend from the floor to the ceiling, bringing natural light to the indoor space.
- The roofs have walkways that invite children to a different atmospheric space.

Hazelwood School Glasgow (Scotland)

- Hazelwood School in Glasgow, Scotland, is a school for children and young people with sensory impairment and complex learning needs.
- The building is designed to maximize natural light and incorporate visual, auditory, and tactile clues.
- The curvilinear building has cork tiles on interior walls that guide children throughout the building.

The Garden School (United Kingdom)

- Located in Hackney, United Kingdom, The Garden School is an outstanding school for 4–16-year-olds with special educational needs, particularly autism.
- The learning space has textured carpets with varying pile heights and wallpapers reminiscent of woodland with multiple trees painted, providing tactile and visual references to nature. These can be used to reduce stress, energize, and relax children and are particularly important for those with special educational needs.
- The playful built-in hexagonal seating areas and spaces serve as small shelters for children to relax and restore their physical and mental energy.

Image Source: Dezeen https://www.dezeen.com/2013/09/09/school-complex-in-rillieux-la-pape-by-tectoniques/

3 CHAPTER BIOPHILIC DESIGN

Image Source: Hazelwood Aerial—Showcase: 4th screenshot, Alan Dunlop Architect, http://www.alandunloparchitects.com/

WeWork's Microschool (United States)

- WeWork created its first microschool in Manhattan, New York, USA.
- The space includes many elliptical objects, fostering a learning environment that is both structured and free-flowing.[57]
- It encompasses modular classrooms, tree houses, a vertical farm, and uses natural materials and neutral colors.

Image Source: Oliver Heath Design
https://www.oliverheath.com/case-studies/the-garden-school/

3 CHAPTER BIOPHILIC DESIGN

Image Source: World Architecture Community https://worldarchitecture.org/architecture-news/ephcg/big-completes-weworks-first-microschool-with-superelliptic-objects-in-new-york-city.html

Climbing-Frame Library (Vietnam)

- Climbing-Frame Library is a library in Hanoi, Vietnam, that has a thriving aquaponics system.
- It is made with a large wooden climbing-frame with concrete stepping stones that children could use to climb, read books, and interact with each other.
- The library incorporates solar-powered aquaponics* to keep vegetables and plants, koi carp, and chickens in the area. The chickens also contribute to the library's small ecosystem, laying eggs that children and other community members could eat, and they provide manure to fertilize the vegetables.[58]
- Through this library, children can learn about self-sustaining ecosystems.

---

* Aquaponics combines the growing of fish with hydroponics—growing plants in water—to make use of the wastewater that fish produce as a nutritious resource for plants and the plant's ability to purify water to provide fresh water for fishes.[59]

3 CHAPTER BIOPHILIC DESIGN

Image Source: Dezeen
https://www.dezeen.com/2019/01/20/vac-library-farming-architects-hanoi/#

How Do You Incorporate Biophilic Design into Learning Environments?

Reflect and Check-in

Before diving into implementing a biophilic design in the space, it is important to reflect on the feelings and concerns regarding this type of design as an individual as well as a community. Below are some questions to consider:

- How do I feel in this space?
- What does my body feel? Do I feel pressure in my chest? Tightness in my throat?
- What can I hear? Smell?
- What does my skin feel?
- What emotions come up as I am in this space?
- What am I reminded of?
- How am I breathing? Am I taking deep breaths? Or are they shallow, slow, or rapid?

Assess the Space

The checklist below helps assess what qualities of biophilic design the space includes:[60]

Nature in the Space
- Views of nature: What nature can I see?
- Sounds, tastes, touch, and smells of nature: What nature can I taste, touch, smell, and/or hear?
- Nonrhythmic sensory stimuli: Can I gaze up at clouds and/or hear bird songs?
- Variations of airflow and temperature: Can I control my sensations of warmth and coolness? Can I feel a breeze?
- Access to water: Are there bodies of water nearby that are clean and natural?
- Access to daylight: Can I see shadows and sun? Can I tell whether it is morning, midday, or evening?
- Seasonal changes and natural cycles of life: Can I grow plants and care for animals? Do I know what the weather is like?

Natural Analogs
- Natural forms: Are there curves and other shapes from nature in the environment?
- Natural materials: Can I identify what things are made of?
- Complexity and order: Is the environment engaging and restorative as opposed to boring or stressful?

Nature of the Space
- Prospect: Can I see far away?
- Refuge: Is there a space I can retreat to?
- Mystery: What entices me to explore the environment?
- The experience of risk with an element of safety: What feels a little scary in the environment?

Analyze the Space

What strengths, weaknesses, opportunities, and "threats" does the space have?

- **Strengths:** Which patterns of biophilic design are visible in the learning space?
- **Weaknesses:** What aspects of your space are not biophilic?
- **Opportunities:** What opportunities do you have for adding biophilic design elements to the learning space? If a parent owns a gardening shop, they may be able to donate plants.
- **Threats:** What external systems, beliefs, and/or processes are preventing you from adding elements of biophilic design to your space?

Once you have analyzed the weaknesses and threats of the space, it is worth brainstorming how you might turn them into strengths and opportunities. For example, if you have beige walls that are not biophilic, you could also see that as an opportunity to apply additional colors to make them biophilic. If someone in your community is concerned about including plants in the space, consider it as an opportunity to communicate, and get to know them and their concerns.

Ideas for Implementing Biophilic Design

Biophilic design does not always mean incorporating actual and simulated views of nature into the space. Nature represented in patterns, fishes, and objects that have biomorphic forms and fractals can also be a representation of nature. Below are some ideas for ways to incorporate biophilic design in your space: [61]

- *Go outside!*
  Take classes outside whenever possible, and try to make the time spent indoors and outdoors equal.
- *Increase natural light*
  Open blinds, turn off artificial lights, add individual task lights.
- *Bring in natural air*
  Open doors and windows whenever possible, introduce fans, take small frequent breaks outside, practice breathing through the nose.
- *Optimize space*
  Have less furniture in the space, de–clutter, introduce furniture with curves and a variety of textures, and bring in fractal images.
- *Create caves*
  Place pillows under tables, create cozy reading and studying nooks, and arrange quiet spaces to sit and think.
- *Place perches*
  Add bar stools and café tables around the perimeter of spaces, position them near a view of nature or nature-based artworks.
- *Add color*
  Add colors that are regional and culturally relevant to nature, hang nature-based artworks, and paint murals.
- *Invite water into the space*
  Add a mini-fountain, go outside in the rain, provide access to fresh clean water, build a pond or a rain garden, and play sounds of water.
- *Engage children in learning about biophilic design and designing indoor and outdoor learning spaces together!*

Additional ideas for implementing biophilic design in learning spaces are located in the appendix.

Get Feedback and Make Changes

Multiple cycles of trial and error make the success of implementing biophilic design possible. Take the time to reflect and get feedback on the space. Below are some questions you could ask:

- What worked?
- What didn't work?
- What would you do differently next time?
- What ideas do you have? What would you share with others about biophilic design? How would you share them?
- What questions do you have?
- What would you want to try next?
- Who else might want to help/get involved?

The Equity Approach: How Does Biophilic Design Contribute to Equitable Education?

 Equitable access to public green spaces is a critical component of both social and environmental justice.
—Emma Urofsky and Robbie M. Parks

As the world becomes more urbanized, many children around the world are living in environments of poor quality due to pollution and lack of access to natural environments. Moreover, many children of color grow up in neighborhoods that have limited access to nature due to systematic racism.

By integrating nature into school designs, schools will ensure children have experience in and build relationships with nature through their everyday learning, no matter their background or identities. Any child in school buildings can benefit from the biophilic design outlined in this white paper. Furthermore, by nurturing children's biophilia through biophilic design, schools can teach the importance of nature and be in an environment that incorporates such elements. This will allow every child to make positive choices that improve their well-being and health as they move on to selecting their own environments.

 Decades of systematic racism have left many people of color living in areas without access to nature.
—Alejandra Borunda

CONCLUSION

Lockdowns during COVID-19 disconnected many people around the world from nature due to architectural practices that failed to make such connections possible. As a result, the integration of nature into the built environment is celebrated even more. Biophilic design provides various benefits to people's health and well-being, and incorporating it into schools carries great opportunities for positively impacting generations of children and removing barriers to learning. When children learn in environments that promote holistic well-being and health, they can experience less stress and anxiety and create meaningful relationships. What is needed in schools as they prepare children for various challenges of the 21st century is to ensure an environment where each and every child could meet the needs of their mind, body, and spirit as it serves as the foundation of their growth and learning.

APPENDIX

Additional ideas for implementing biophilic design in learning spaces:
Link to table: *Biophilic Design in Learning Spaces* at: https://tinyurl.com/367ppbkk

Source: "An Introduction to Biophilic Design. Interface." (2022).
https://www.interface.com/AU/en-AU/design/biophilic-design.html

ENDNOTES

1. Oliver Heath, "Human Spaces: Creating Positive Spaces Using Biophilic Design." Interface, June 2018. https://blog.interface.com/en-au/creating-positive-spaces-using-biophilic-design/.

2. Weijie Zhong et al., "Biophilic Design in Architecture and Its Contributions to Health, Well-Being, and Sustainability: A Critical Review." Frontiers of Architectural Research (2021). https://doi.org/10.1016/j.foar.2021.07.006.

3. Heath, "Human Spaces," (n 1).

4. Anna Klare Harrison, "Wonder and Awe: How Biophilic School Design Removes Barriers to Learning," [Video], YouTube, May 19, 2020. https://www.youtube.com/watch?v=tAjmfUSWuGc&ab_channel=AnnaKlareHarrison

5. Terrapin Bright Green, "14 Patterns of Biophilic Design," September 12, 2014, https://www.terrapinbrightgreen.com/reports/14-patterns/#rediscovering-the-intuitively-obvious.

6. Florence Williams, "The Nature Fix: Why Nature Makes Us Happier, Healthier, and More Creative," 2016, http://www.florencewilliams.com/the-nature-fix.

7. Terrapin Bright Green, "14 Patterns," (n 5).

8. Harrison, "Wonder and Awe," (n 4).

9. Jim Determan et al., "The Impact of Biophilic Learning Spaces on Student Success." Architecture Planning Interiors, 2019. https://www.brikbase.org/content/impact-biophilic-learning-spaces-student-success

10. Rokhshid Ghaziani, School Design with Children: Promoting Mental Health and Wellness in Schools by Incorporating Biophilic Design, (De Montfort University, February 24, 2020), PDF. https://associatedevelopmentsolutions.com/wp-content/uploads/2020/03/biophilic-design-for-schools-rokhshid.pdf

11. Harrison, "Wonder and Awe," (n 4).

12. Randy White and Vicki L. Stoecklin, "Nurturing Children's Biophilia: Developmentally Appropriate Environmental Education for Young Children." Collage: Resources for Early Childhood Educators, (November 8, 2008): 1–11. https://www.whitehutchinson.com/children/articles/nurturing.shtml.

13. Ibid.

14. Interface, "An Introduction to Biophilic Design," 2022, https://www.interface.com/AU/en-AU/design/biophilic-design.html.

15. Determan, "The Impact of Biophilic Learning," (n 9).

16. Ibid.

17. Ibid.

18. National Institute of General Medical Sciences. "Circadian Rhythms," May 4, 2022, accessed July 26, 2022, https://www.nigms.nih.gov/education/fact-sheets/Pages/circadian-rhythms.aspx.

19. Determan, "The Impact of Biophilic Learning," (n 9).

20. Ibid.

21. Ibid.

22. World Health Organization. "Adolescent Mental Health," November 17, 2021, accessed August 10, 2022, https://www.who.int/news-room/fact-sheets/detail/adolescent-mental-health#:%7E:text=Key%20facts,illness%20and%20disability%20among%20adolescents.

23. Determan, "The Impact of Biophilic Learning," (n 9).

24. Harrison, "Wonder and Awe," (n 4).

25. Determan, "The Impact of Biophilic Learning," (n 9).

26. Ibid.

27. Ibid.

28. Ibid.

29. Ibid.

30. Terrapin Bright Green, "14 Patterns," (n 5).

31. Ibid.

32. Ibid.

33. Ibid.

34. Ibid.

35. Harrison, "Wonder and Awe," (n 4).

36. Terrapin Bright Green, "14 Patterns," (n 5).

37. Ibid.

38. Ibid.

39. Ibid.

40. Ibid.

41. Ibid.

42. Ibid.

43. Harrison, "Wonder and Awe," (n 4).

44. Heath, "Human Spaces," (n 1).

45. Terrapin Bright Green, "14 Patterns," (n 5).

46. Ibid.

47. Harrison, "Wonder and Awe," (n 4).

48. Terrapin Bright Green, "14 Patterns," (n 5).

49. Ibid.

50. Ibid.

51. Ibid.

52. Ibid.

53. Ibid.

54. Ibid.

55. Ibid.

56. Ibid.

57. Ghaziani (n 8).

58. Jon Astbury, "Climbing-Frame Library in Vietnam Has a Thriving Aquaponics System," Dezeen, February 12, 2022, https://www.dezeen.com/2019/01/20/vac-library-farming-architects-hanoi/#.

59. Ibid.

60. Anna Klare Harrison, Biophilic Design for Authentic Deeper Learning, [Slides] 2021. https://drive.google.com/file/d/1txO7tg5GSw3JG8WUt53Ml6WxRvCvkvR0/view.

61. Ibid.

3 CHAPTER BIOPHILIC DESIGN

A Pictorial Essay

An important design approach to biophilic design is to start with natural materials and forms that are more closely associated with nature and natural settings such as this new barn which houses the Bowers School Farm in Michigan. This 12,000 sq. ft. facility uses geothermal heating, septic fields, a bio-retention pond and a fully operational greenhouse. Photo by Christopher Lark.

3 CHAPTER BIOPHILIC DESIGN

The best way for design to express a love for nature (biophilia) is to take learning outside as with these decks at Hillel Academy, Tampa that flow naturally from indoor learning areas.

3 CHAPTER BIOPHILIC DESIGN

This veranda adjacent to the art room at DSB International School in Mumbai set in a dense urban area shows that biophilia can be expressed almost anywhere with some imaginative design.

In settings where the weather permits, there is almost nothing that cannot be learned better outside than in indoor settings far removed from daylight and fresh air.

A green amphitheater like this one at Swarthmore College is an elegant way to express a biophilic design solution since it can permit social distancing and also allow for many more modes of learning such as music, dance and performance, student presentations and quiet reading. It is also a healthy, comfortable and inspiring space with lots of fresh air. Compare this to a traditional indoor classroom!
Photo: *Swarthmore College.*

This is a generously sized vegetable garden at Learning Gate Community School in Lutz, Florida where students do most of the planting and tending. A substantial part of every student's day at this school is spent outdoors, and this was true even before the COVID-19 pandemic.

Biophilic design becomes the means by which school courtyards such as this one at Riverside School in Ahmedabad, India feel warm, welcoming, and home-like via the use of indigenous plantings and an abundance of greenery.

Collingwood College in Melbourne, Australia, pioneered the Stephanie Alexander Kitchen Garden Foundation project where children have the opportunity to plant, grow, harvest, cook, and eat organic foods.

3 CHAPTER BIOPHILIC DESIGN

Good biophilic design does not have to be expensive as seen by this modest renovation in which a simple wood trellis provides shade and a quiet place to study between two buildings.

3 CHAPTER BIOPHILIC DESIGN

Play Inspired by Biophilia: An unstructured area with an assortment of natural elements such as the one featured in this photo inspires more creative play than a structured playground does.

A perfect example of biophilic design is one where water is naturally recycled as with this greenhouse with a living machine at Oberlin College, Ohio. (Photo couesy of Barney Taxel).

3 CHAPTER BIOPHILIC DESIGN

While its easier to build on a site where all existing trees are first removed, preserving and building around a large tree can immediately introduce nature within the building in a powerful way, bringing with it all the ancillary health and learning benefits of biophilic design.

3 CHAPTER BIOPHILIC DESIGN

Balconies and verandas that have good nature views and that can be easily accessed from interior learning spaces are easy and relatively inexpensive examples of biophilic design.

3 CHAPTER BIOPHILIC DESIGN

At the AMIT GOGYA teacher training academy in Israel, architect Roni Zimmer Doctori used this warm wood-finished deck in place of the deep concrete "amphitheater" that existed here before. Human beings are naturally drawn to materials and finishes like wood and bamboo over more institutional materials like steel and concrete. biophilic design will therefore prefer natural surfaces and finishes whenever possible.

When it is not possible for students to be out in nature, then it makes sense for nature to be brought indoors as with this large green indoor planter. Such installations are even more critical in cold climates where students live are deprived of green settings in the outdoors for several months of the year.

3 CHAPTER BIOPHILIC DESIGN

While green amphitheaters are the most desirable, more modest ones like these paved examples can also work if they are adequately shaded. Natural stone can qualify as appropriate biophilic design and will be even more effective when used in a green setting with trees all around as with the paved amphitheater below at the American Embassy School in New Delhi, India.

3 CHAPTER BIOPHILIC DESIGN

Water elements inside and outside a school building are often powerful examples of good biophilic design as demonstrated by this Koi pond at Sinarmas World Academy in Jakarta, Indonesia.

For biophilic design to be effective in outdoor settings, it is important to plan properly for the kinds of activities that will happen there. Table surfaces that can be easily cleaned are more suitable for outdoor art projects such as the one from Israel shown in this photo.

3 CHAPTER BIOPHILIC DESIGN

At Anne Frank Inspirer Academy in San Antonio, the trees that were cut down on the site to make room for the new buildings were repurposed as part of a feature wall (background). The wood-finished ceiling, natural wood floors, and wooden cubbies are all set in a brightly daylit space with nature views. These elements combine to create an environment that is quintessential example of holistic biophilic design.

3 CHAPTER BIOPHILIC DESIGN

Dr. Parul Minhas
Prakash Nair, AIA

SALUTOGENIC DESIGN

To Promote Student Health and Well-Being

 Salutogenesis aims to prevent disease and promote health.

Introduction

The U.N. World Health Organization defines health (1948) as a "state of complete physical, mental, and social well-being and not merely the absence of disease or infirmity". This holistic concept of health recognises the impact of social, economic, political, and environmental influences on health involving the well-being of the whole person. It is also critical to note that it is within human capacity to influence and modify the environment (Morandi et al., 2011).

The notion of holistic well-being cuts across cultures. Ayurveda (one of the world's oldest holistic healing systems) describes health as 'swasthya'; a Sanskrit term meaning 'stability in the true self' that is a state of complete, balanced, physical, mental, and spiritual well-being (Sharma et al., 2007). This approach of Ayurveda is in sync with WHO's definition of health as mentioned earlier.

The concept of salutogenesis, as described by Antonovsky (1979), has a striking resemblance and correlation with the abovementioned approaches towards health (Shivam S. Gupta and Satyam S. Gupta, 2019). Ayurveda and salutogenesis are both interculturally and universally applicable beyond cultural or ethnic backgrounds (Morandi et al., 2011). Ayurveda goes into depth, revealing the inner coherence of the system through observation (Morandi et al., 2011).

Like Ayurveda, salutogenesis aims to prevent disease and promote health. "Health promotion is the process which enables people to gain control over their health determinants in order to improve their health and thereby be able to live an active and productive life" (WHO, 1986). In a global world, characterised by rapid social changes, the ability to manage stress that is related to change is crucial for the maintenance and development of health and quality of life.

Holistic health may, therefore, be defined as the ability to maintain a state of equilibrium and balance between genetic factors and environmental conditions, mental-spiritual and bodily functions along with the interaction between individual and community together, leading to the attainment of full human potential (self-actualisation) and building of a sound coping mechanism (resilience).

 The notion of holistic well-being cuts across cultures.

4 CHAPTER SALUTOGENIC DESIGN

FIGURE 57. *A model of holistic health with self-actualisation and resilience as the ultimate goals to be achieved through mind-body-spirit balance in the presence and under the influence of physical and social environmental conditions. From Parul Minhas, created in Word, 11-21-21.*

This paper is primarily based on theory of salutogenic design that aims to promote holistic health in learning environments with an intention of guiding architects and school planners in the design process. Salutogenic design theory has been derived from Antonovsky's studies and, specifically, his in-depth study of environmental factors that promote health. His key conclusion was that relatively unstressed people had much more resistance to illness than those who were more stressed.

Antonovsky, who coined the term salutogenesis, proposed that the experience of well-being is based on a 'sense of coherence' (SOC). Sense of coherence is a pervasive, long-lasting, and dynamic feeling of confidence that one's internal and external environments are predictable and that there is a high probability that things will work out as well as can be expected (Antonovsky, 1979). Research carried out in the past few decades has confirmed that SOC has strong positive correlations to perceived health, mental health, and quality of life. The SOC has three components:

i) Comprehensibility based on cognition ("My world is understandable")
ii) Manageability based on coping ability ("My world is manageable")
iii) Meaningfulness, which gives motivation ("My world has meaning").

Thus, an environment that helps in enhancing these three components of SOC can be classified as a salutogenic or a health promoting environment.

 Sense of coherence is a pervasive, long-lasting, and dynamic feeling of confidence that one's internal and external environments are predictable

Child Health and Built School Environment

Because children spend nearly half of their waking hours in school, the school environment plays a critical role in a child's holistic health. "Health problems developed in young children typically affect the child's social, behavioral, cognitive, and physical processes and have the tendency to be compounded through aging. How the child contends with internal health factors, external environmental factors and issues of self-identity play a key role in holistic health development." (Hembree & Sholder, 2013). This study calls for the development of a child's holistic health by recognizing the needs of the mind, body, and spirit. These three pillars of good health should be addressed in any building designed to serve children and help enhance a child's self-image.

In order to frame salutogenic design guidelines for learning environments, it is important to decipher the holistic health needs of children that must be met in their social and physical environment in order for them to attain a high SOC. To determine this, a variety of studies in children's health, child psychology, and environmental psychology were consulted, including Maslow's hierarchy of needs (1943), India Child Well-Being Report (2019), self-determination theory by Deci & Ryan (1985), and / Hughes, Willis, & Franz (2015).

Maslow's hierarchy of needs (Maslow, 1943) states that people must have their basic physiological and psychological needs met to pave way for higher goals like self-actualisation. Deci & Ryan (1985) stated, "Conceptualized basic psychological needs for autonomy, competence, and feeling connected as innate and must be fulfilled for overall health and well-being."

TABLE 4. What Children 'Need' to be Healthy?

Holistic Health Needs Of Children
Habitable environment (Clean air, water, shelter, thermal comfort, natural light, etc.)
Safety & Security
Self-esteem/ Personal growth/ Self-acceptance
Autonomy/ psychological freedom
Positive relationships with people and places
Rich experiences leading to positive emotions/Joy/Empathy/Compassion/Enthusiasm
Competence/ Capability/ Accomplishment/ Mastery
Engagement/ Purpose in life

What children need to be healthy? From Parul Minhas (2021)

4 CHAPTER SALUTOGENIC DESIGN

Built School Environment

Spatial cues from the envisioned learning environment of some renowned educators, architects, and psychologists can greatly help to decipher the design considerations of a healthy school design. Table 5 mentions the prominent keywords from their education design philosophy:

TABLE 5. Attributes of an Ideal School Environment.

Educator/Architect/Psychologist	Spatial cues from their envisioned learning environment for children
Mahatma Gandhi (Singh, 2019)	Naturalistic setting, flexible spaces, autonomy, sense of belonging, experiential learning, self-esteem, holistic approach
Rabindra Nath Tagore (Tirath, 2017)	Connect between man and nature, autonomy, harmony with environment, holistic approach
Krishnamurthi (Lichtenberg, 2010)	Connect with nature, space for self-introspection, sense of belonging
Montessori (Gutek, 2004)	Self-exploration, interaction with environment, flexible spaces, relevance of second plane of development (6–12 years)
John Dewey (1907)	School as a miniature community, social interaction
Froebel (Roszak, 2018)	Freedom with guidance, social interaction, connection with nature and space for self-activity and reflection
Rudolf Steiner (Bjørnholt, 2014)	Influence of aesthetics and architectural forms, developmental stages
Piaget (1969)	Developmental stages, interaction with environment, sense of ownership
Lev Vygotsky (Ivic, 1994)	Social interaction, mixing of age groups
Loris Malaguzzi (Ellis, 2007)	Socialisation, interaction with environment, natural light, multiple modes of learning
Urie Bronfenbrenner (Darragh, 2006)	Stimulation of senses, flexibility, interaction with environment
Carl Rogers (Morgan, 1977)	Positive school climate, autonomy, holistic approach
Herman Hertzberger (2008)	Stimulating environment, social interaction, flexible spaces,
Howard Gardner (1993)	Self-esteem, flexibility to accommodate multiple modes of learning
Mark Dudek (2008)	Autonomy, social interaction, school as community
Henry Sanoff (2001)	Sense of belonging, aesthetics, community participation
Christopher Day (2007)	Self-esteem, stages of development, sense of belonging
Nair & Fielding (2009)	School as a small learning community
Walden (2015)	Social interaction, child's (user) perspective, stages of development

Attributes of an Ideal School Environment. From Parul Minhas (2021).

Salutogenic Design and the School Environment

Dilani (2008) conceived the idea of salutogenic design, or what he called "Psychosocially Supportive Design" to promote health. He maintained that salutogenic design not only defined the causes of stress but also introduced wellness factors that could strengthen health. The theory suggested that salutogenic design considerations could help create designs that not only reduced stress but focused on salutary (health promoting) factors.

Salutogenic design in an educational context must aim at identifying the elements of physical school design that can contribute towards the development of a strong SOC, leading to improved holistic health of children. Dilani (2008) created a list of architectural characteristics that he argued could strengthen an individual's SOC.

Addressing the SOC in the context of school design, we can elaborate on the three attributes of SOC in **Figure 58**.

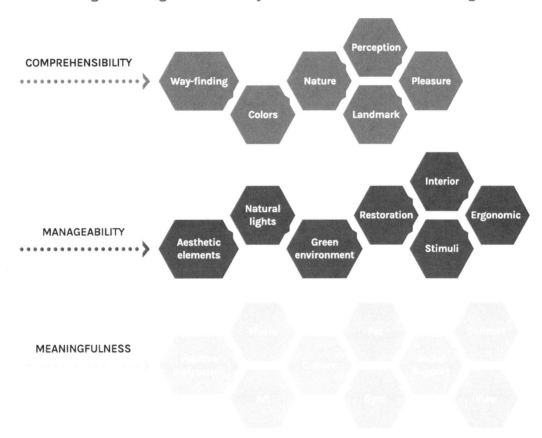

FIGURE 58. Design factors in relation to sense of coherence. Adapted from "Psychosocially Supportive Design: A Salutogenic Approach to the Design of the Physical Environment," by Alan Dilani, 2008, Design and Health Scientific Review, 1(2), pp. 47–55. (https://www.researchgate.net/publication/265349464_Psychosocially_Supportive_Design_A_Salutogenic_Approach_to_the_Design_of_the_Physical_Environment)

 Salutogenic design considerations could help create designs that not only reduced stress but focused on salutary (health promoting) factors.

Comprehensible School Environment

According to Krause (2011), experiences of consistency are the basis for the development of comprehensibility. In positive cases, children have feelings of security and acceptance in social relations. Consistency in experiences comes when most events in daily lives are predictable. Though it is neither possible nor desirable to predict every experience as it may lead to monotony, human beings flourish when most of their experiences are consistent so that they can spare more time to pursue what they want to rather than adjusting to unpredictable events/experiences. When translated to a built school environment, experiencing consistency would mean being able to comprehend the connection between the various spaces and having confidence that they all connect to form a unified whole, leading to a sense of security and coherence. A secure environment must, therefore, possess the qualities of being decipherable (Day, 2007) and transparent (Nair, Fielding, & Lackney, 2009). These environments orient and reassure children by using familiar elements and special features that may assist wayfinding and legibility (Dilani, 2008). It requires an optimum organization of space to control density and assure personal space for everyone. Comprehensible environments are authentic, genuine, and honest, and these qualities may be conveyed through the use of natural materials and construction methods, usually avoiding superfluous decoration and detailing (Hughes, Wills, & Franz, 2019). According to Ken Yeang (mentioned in Hughes, Willis, & Franz, 2019), '"environmental comprehensibility" requires environmental orderliness, predictability, and legibility. This includes, for instance, the importance of visual order in the built environment with legibility, intuitive wayfinding and elimination of visual chaos, etc.' (Dilani, 2015). The following design cues for the derivation of 'comprehensible' design guidelines for a salutogenic learning environment are listed by combining the holistic health needs (Table 1) and attributes of ideal learning environment (Table 2) with comprehensibility considerations as previously discussed.

 These environments orient and reassure children by using familiar elements and special features that may assist wayfinding and legibility

Design Cues for a Comprehensible School Design

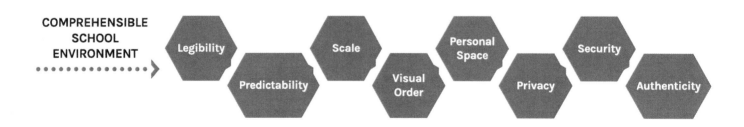

FIGURE 59. *Design Cues for a Comprehensible School Design Source: Parul Minhas, The Design of Learning Environments to Promote Student Health & Well-Being (PhD thesis, 2021).*

4 CHAPTER SALUTOGENIC DESIGN

Manageable School Environment

Krause (2011) insisted that experiences of self-efficacy are the basis for the development of manageability. This component grows if the requirements for children are available to their developmental level and if they experience the acceptance of their progress. According to Hughes, Wills, & Franz (2019), manageable school environments are those that aim to build competence by being well resourced, enhancing the ability to cope, develop further capabilities, and undertake required/desired activities. These resources could also be the environments that allow students to exercise control and support activities by being safe, comfortable, and accessible. Inclusive design also forms a part of a manageable environment where students with special needs are taken into consideration. Research on inclusive and universal design provides further support (Myerson & Lee, 2010; Khare, Mullick, & Raheja, 2011). It is also crucial for a manageable environment to be flexible and responsive to change and to encourage participatory planning. Dilani (2008) suggested that environmental components that foster manageability are aesthetics, natural light, green environments, restoration, stimuli, and ergonomics. Comprehensibility is a pre-condition for effective manageability. The following design cues for the derivation of 'manageable' design guidelines for a salutogenic learning environment are listed by combining the holistic health needs (Table 5) and attributes of ideal learning environment (Table 6) with manageability considerations discussed above.

 It is also crucial for a manageable environment to be flexible and responsive to change and to encourage participatory planning.

Design Cues for a Manageable School Design

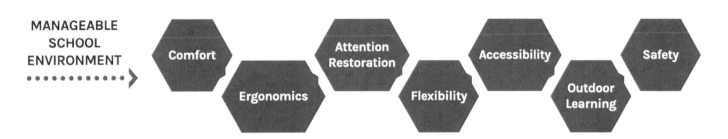

FIGURE 60. *From Parul Minhas The Design of Learning Environments to Promote Student Health & Well-Being (PhD thesis, 2021).*

Meaningful School Environments

Krause (2011) observed that the motivational and emotional component increases when children can influence and participate in social decision-making processes (sense of purpose). Children need to feel that they belong to the school and school belongs to them. Based on research data from neurobiology and resilience studies, it can be stated that experiencing a minimal amount of empathic resonance is a fundamental biological need without which the human being could not survive (Krause, 2011). If a child feels accepted and acknowledged, he/she feels recognized and gets feedback, which strengthens the self-worth. According to Hughes, Willis, & Franz (2019), an environment that motivates children's desire for a SOC is perceived to be meaningful. Such environments are 'inspiring, engaging, restoring, challenging, and aesthetically rich' (Krause, 2011). Natural and built environments that engage the senses through material qualities of 'color, texture and pattern' and through atmospheric qualities of 'light, temperature and sound' are particularly important in this context. Alongside natural elements, several other additions can make an environment meaningful—for example: music, art, culture, gym, spaces for social support, opportunity to interact with other species i.e., pets and other positive distractions (Dilani, 2008).

A meaningful environment must, therefore, be able to evoke feelings of belonging (self-worth) and engage people positively so that they experience a sense of purpose. The following design cues for the derivation of 'meaningful' design guidelines for a salutogenic learning environment are listed by combining the holistic health needs (Table 5) and attributes of an ideal learning environment (Table 6) with meaningfulness considerations as previously discussed.

Design Cues for a Meaningful School Design Source

FIGURE 61. *Design cues for a Meaningful School Design From Parul Minhas, The Design of Learning Environments to Promote Student Health & Well-Being (PhD thesis, 2021).*

Twenty-Seven Design Guidelines for Salutogenic Learning Environments

The three components of salutogenic design, as discussed above, are not mutually exclusive but closely knit and interdependent. Each one of the following guidelines may help in enhancing one or more components of salutogenic learning environment. These guidelines are a broad and generalized set of instructions and are intended to guide the architects through the design process. The guidelines must be contextualized before their application. Physical design of a facility for various age groups (e.g., Piaget's stages of development) may vary considerably under the same design guideline. Similarly, the variations in school size, budget constraints, cultural values, and other factors may lead to a variable physical manifestation of the same guideline. The salutogenic approach of health promotion, however, remains the same.

The guidelines in Table 7 are for salutogenic learning environments and can also be categorized as macro (1–14) and micro (15–30). The impact of macro guidelines is more visible and direct while the impact of micro guidelines is more subtle and experiential. However, they are equally relevant.

TABLE 6. *Twenty-Seven Design Guidelines for Salutogenic (Health Promoting) Learning Environments.*

Twenty-Seven Design Guidelines for Salutogenic (Health Promoting) Learning Environments
1. Use effective wayfinding strategies to improve legibility and build assurance.
2. Ensure safe community involvement and access control for enhancing competence.
3. Consider appropriate scale & developmental needs to foster autonomy and self-esteem.
4. Apply ergonomic considerations to improve posture and increase efficiency.
5. Use colors carefully to avoid visual fatigue and enhance psychological comfort.
6. Provide outdoor spaces to augment collateral learning and social connections.
7. Manage density & crowding to improve self-worth and ensure social distance.
8. Apply universal design principles to improve accessibility and build self-esteem.
9. Ensure ample natural light for enhancing overall health and efficiency.
10. Facilitate multiple modalities of learning by designing flexible spaces.
11. Provide common spaces for peer/ teacher interaction to enhance a sense of community/belonging.
12. Provide a variety of spaces that enhance engagement and initiate a state of flow.
13. Ensure effective noise reduction to combat stress and improve efficiency.
14. Maximize natural ventilation and thermal comfort to improve efficiency and overall health.
15. Create a welcoming entry and signature elements to help children comprehend the environment better and feel connected.
16. Ensure safety and security through natural surveillance and other design interventions.
17. Allow personalization of spaces to encourage ownership/territoriality
18. Use sensory aesthetics and active design elements for optimum stimulation (physical & psychological).
19. Provide spaces for self-reflection (privacy) and small group activities to enhance autonomy, competence, and relatedness.
20. Maximize authenticity and psychological comfort through the use of appropriate materials.
21. Create homelike environments with familiar elements to enhance the feeling of security and psychological freedom.
22. Promote agentic learning and environmental stewardship through visible green/sustainable architecture features.
23. Apply biophilic design principles to counter nature deficit and for effective attention restoration.
24. Provide spaces for pets in school in order to inculcate responsibility and empathy.
25. Allow student participation in planning and design of new facilities as well as maintenance/renovation projects.
26. Provide stimulating playgrounds to build risk competence and to experience a sense of adventure.
27. Enhance overall sense of coherence (SOC) by splitting bigger schools into small learning communities.

Source: Parul Minhas The Design of Learning Environments to Promote Student Health & Well-Being (PhD thesis, 2021).

4 CHAPTER SALUTOGENIC DESIGN

Healthy School Assessment Tool

 The score under each category can rightly direct the resources of renovation projects that aim towards health promotion of schoolchildren.

The guidelines can greatly help in the creation of new learning environments, but it is equally significant to create an assessment tool for conducting a post occupancy evaluation (POE) to determine the health status of existing facilities in order to make the required interventions wherever required. For this purpose, a school assessment tool has been formulated on the basis of the above guidelines. The score under each category can rightly direct the resources of renovation projects that aim towards health promotion of schoolchildren. This checklist is purely for guiding design decisions and does not, however, include building condition assessment that must be carried out separately for renovation projects. The following tool has been approved and validated by school design expert Ar. Prakash Nair (Author of Language of School Design, 2009) and Ar. Alan Dilani (Pioneer of Salutogenic Design). The established school assessment methods by Sanoff (2001) and Tanner (2009) have been the major references for the creation of this tool. The average score in each of the 27 categories indicates the extent to which the school satisfies the given health consideration.

School Details

Name of the school:	
Location/Address:	
Site area:	
Built up area:	
No of students	
No of teachers	
Average class size	
Average no. of students/class	
Other details	

Code assigned

4 CHAPTER SALUTOGENIC DESIGN

Healthy School Assessment Tool (HSAT)

1. Legibility and wayfinding	0	1	2	n/a

1.1 A unique identity is created for each location

1.2 Use of landmarks as visual cues

1.3 Well-structured paths with goals

1.4 Limited navigational choices

1.5 Sightlines are used to show what is ahead

1.6 Covered pathways among buildings within the campus

1.7 Color coded indoor pathways

1.8 Clear and well-lit pathways to activity areas

1.9 The main building has an obvious point of reference among the school's buildings in which paths and buildings connect

1.10 Distinction between various areas is made obvious by the use of colors, textures, forms, ceiling heights, etc.

SCORE

0 = Inadequate, 1 = Adequate, 2 = Excellent

2. Community involvement and access control	0	1	2	n/a

2.1 Well placed windows to get a clear view of the entrance

2.2 Dedicated areas for community interaction near the entrance

2.3 Signage and pavements to define accessible areas for visitors

2.4 Clearly defined limits to ensure access control

SCORE

0 = Inadequate, 1 = Adequate, 2 = Excellent

3. Child scale and developmental needs	0	1	2	n/a

3.1 Variation in ceiling heights acc. to the intended use of space

3.2 Spaces and furniture considering child scale

3.3 Whiteboard and other equipment respecting child scale

3.4 Soft classrooms with curvilinear shapes, pillows, rugs, etc.

3.5 Accessible material storage

3.6 Door handles, switches, etc. at child's scale

3.7 Variety of sizes of spaces

SCORE

0 = Inadequate, 1 = Adequate, 2 = Excellent

4 CHAPTER SALUTOGENIC DESIGN

4. Ergonomic considerations for posture correction	0	1	2	n/a
4.1 Variety of furniture that is flexible and easy to use.				
4.2 The furniture improves posture and is in good repair.				
4.3 Workstations are designed to accommodate information technology.				
4.4 Floor seating and opportunity for reclining provided at a corner of the classroom.				
4.5 Popliteal heights, elbow angle, and other anthropometric considerations are taken care of.				
4.6 Footrest is provided for shorter children.				
4.7 Tables and built-in shelves have rounded edges.				
SCORE				

0 = Inadequate, 1 = Adequate, 2 = Excellent

5. Careful use of color	0	1	2	n/a
5.1 Variety of colors used while being careful about overstimulation				
5.2 Contrast between the board and the back wall just appropriate				
5.3 Color and lighting are considered together				
5.4 Colorful displays on the walls and doors				
5.5 Warmer tones are preferred for younger children and cooler tones for older children				
5.6 Physical activity areas like gyms, yoga halls, etc. are painted in cooler colors				
5.7 Privacy niches and other areas for withdrawal are painted in cooler colors				
5.8 Stage area in auditorium is in contrast with surroundings and is painted in relaxing colors like beige, peach, or pastel green				
SCORE				

0 = Inadequate, 1 = Adequate, 2 = Excellent

6. Outdoor learning spaces	0	1	2	n/a
6.1 Provision of an accessible green/open space immediately outside the classroom				
6.2 An amphitheater readily available for outdoor plays, performances, and presentations.				
6.3 Provision of temporary structures and benches to facilitate outdoor classroom				
SCORE				

0 = Inadequate, 1 = Adequate, 2 = Excellent

7. Density and crowding | 0 | 1 | 2 | n/a

7.1 Ample space to move around in the classroom

7.2 Gross Area provision per child is between 7- 10 sq.m.

7.3 Children divided into smaller groups/cohorts

7.4 Uncluttered rooms as well as walls

7.5 No. of students per class is between 17-25

SCORE

0 = Inadequate, 1 = Adequate, 2 = Excellent

8. Accessibility and universal design | 0 | 1 | 2 | n/a

8.1 Simple, clear circulation with clearly defined paths, doorways, etc.

8.2 Provision of handrails as necessary and material textures considered as tactile means of way finding.

8.3 Power doors to improve accessibility for all users.

8.4 Provision of ramps/lifts for barrier free access

8.5 "Maze" entrances to washrooms improve access for all users and reduce hygiene issues associated with door knobs/levers.

8.6 Circulation routes are of appropriate width (min. 1.5m wheelchair turning diameter) and are kept clear of obstacles.

8.7 Hardware and controls are located within reach of users and ensure ease of operation.

8.8 Special consideration of acoustics for the visually impaired: buildings and rooms are designed to reduce echo, reverberation, and extraneous background noise.

8.9 Provision of appropriate lighting (natural and artificial) for circulation. Glare is avoided though.

8.10 Large flat panel light switches, which can be used with either hand, closed fist, elbow, etc. are provided.

8.11 Rough or textured borders, which contrast with smooth walking surfaces and indicate a change in grade or material, are used.

8.12 Door lever does not require grip strength and can be operated by a closed fist or elbow.

SCORE

0 = Inadequate, 1 = Adequate, 2 = Excellent

4 CHAPTER SALUTOGENIC DESIGN

9. Natural light for overall health and efficiency	0	1	2	n/a
9.1 Diffused (glare-free), usable daylight in every space where children spend long periods of time.				
9.2 Smaller windows at eye level are installed for views along with skylights or clerestory windows high in the wall deliver glare-free light deep into the space.				
9.3 Unrestricted views (when glare is not a problem) provide a perspective to ease eyestrain and bring the outside and inside worlds together.				
9.4 Daylight is supplemented with electric light. An acceptable design includes artificial light plus natural light from the outside.				
9.5 Direct view of bright light sources like the sun, a bright sky, or an electric lamp that may create glare and visual discomfort is avoided.				
9.6 Diffused daylight enters from multiple directions and minimizes shadows, balancing the light across the room.				
9.7 For every 10 square meter of classroom floor space, at least 2.5 square meter of window space is provided.				
9.8 Windows have some form of glare control, but are in use (when glare is not a problem), and are without painted obstructions.				
SCORE				

0 = Inadequate, 1 = Adequate, 2 = Excellent

10. Flexible spaces to facilitate multiple modalities of learning	0	1	2	n/a
10.1 Space can be made larger/ smaller or of a varying shape with a few changes in furniture arrangement				
10.2 The spatial layout allows the use of multiple learning modalities				
10.3 Movable and flexible partitions that can be operated easily				
10.4 Possibility for expansion/change is present				
10.5 Adjustable furniture to support both technology use and writing/drawing, etc. by hand				
10.6 Curtains/blinds, etc. to allow the usage of projector, SMART Board®, etc.				
SCORE				

0 = Inadequate, 1 = Adequate, 2 = Excellent

11. Variety of engaging spaces that initiate a state of flow	0	1	2	n/a
11.1 Engaging library with vibrant furniture, furnishings, colors, etc.				
11.2 Reading areas are well-lit with spaces for group work				
11.3 Acoustically well designed and well-lit music and dance areas				
SCORE				

0 = Inadequate, 1 = Adequate, 2 = Excellent

4 CHAPTER SALUTOGENIC DESIGN

12. Common spaces for peer/teacher interaction	0	1	2	n/a

12.1 Enough space/opportunities for 4 or more children, in more than 3 locations in visible/safe locations is provided

12.2 Space for having a meal together with peers and teachers

12.3 Presence of auditorium, amphitheater, etc. in school

SCORE

0 = Inadequate, 1 = Adequate, 2 = Excellent

13. Natural ventilation and thermal comfort	0	1	2	n/a

13.1 Passive techniques for thermal insulation in extreme climates

13.2 Provision for windows at various levels to be used during varied weather conditions

13.3 Less noisy mechanical systems, if any

13.4 Mechanical systems with a capacity to draw significant amount of outside air into the building

13.5 Use of natural airflow patterns to circulate fresh air

13.6 Higher ceiling heights wherever possible

SCORE

0 = Inadequate, 1 = Adequate, 2 = Excellent

14. Effective noise reduction	0	1	2	n/a

14.1 Site located in a peaceful area with low noise and air pollution

14.2 Barriers and buffers are provided to counter traffic noise

14.3 Landscaping is used as a dampener

14.4 Toilets, storerooms, etc. are used as buffer zones

14.5 Hard materials that cause echo are avoided

14.6 Appropriate sound absorbing materials are used

SCORE

0 = Inadequate, 1 = Adequate, 2 = Excellent

4 CHAPTER SALUTOGENIC DESIGN

15. Welcoming entry and signature elements emphasizing the cultural context	0	1	2	n/a
15.1 An inviting and highly visible entrance with well-defined architectural features, signs, lighting, artwork, landscaping, and other landmarks such as flags				
15.2 Scale of the entrance is not intimidating for the child				
15.3 Motivational signs that send positive messages and invite children to school are used				
15.4 Landscaping features or small play areas are visible from the entrance				
15.5 Covered entrance that provides shelter from bad weather and facilitates transition				
15.6 Safe drop off/pick up				
15.7 Separate access for students and visitors				
15.8 Signature elements emphasizing the local/cultural context				
SCORE				

0 = Inadequate, 1 = Adequate, 2 = Excellent

16. Safety and security through natural surveillance	0	1	2	n/a
16.1 Centrally located administrative offices to enhance student safety				
16.2 Parking areas are delineated for staff and visitors				
16.3 Entrances and exits are easily and effectively monitored				
16.4 All student/pedestrian pathways are passively monitored				
16.5 No hiding spaces created by landscaping, fencing, etc.				
16.6 Separate age-level playgrounds for various age cohorts with developmentally appropriate and safe playground equipment				
16.7 Extensive use of windows and glazed doors to enhance natural surveillance of entrances, pathways, etc.				
16.8 No unattractive barriers such as barbed wire on the school grounds				
16.9 Toilets are attached to classrooms, if not then they have auditory connections with adjoining areas				
16.10 Security devices are unimposing				
16.11 Security system (alarms, lights, locks) provides elevated levels of security				
16.12 The site and learning environments are free of excessive non pedestrian traffic, hazards, and noise				
16.13 There are no high voltage power lines in the close proximity of the school				
SCORE				

0 = Inadequate, 1 = Adequate, 2 = Excellent

4 CHAPTER SALUTOGENIC DESIGN

17. Ownership/territoriality through personalization of spaces	0	1	2	n/a

17.1 Personal workspace with lockers for each student

17.2 Spaces for personal artifacts

17.3 Personal storage for books, stationary, etc.

17.4 Distinctive design elements and display spaces for student works

SCORE

0 = Inadequate, 1 = Adequate, 2 = Excellent

18. Sensory aesthetics and active design elements	0	1	2	n/a

18.1 Attractive and plenty of well-maintained landscape areas

18.2 Highly articulated fenestrations for framing of views

18.3 Visually pleasing staircases and other movement pathways to encourage walking

18.4 Age-appropriate design of walking routes

18.5 Avoidance of long narrow corridors and use of nature connected pathways instead

18.6 Provision of dedicated indoor spaces for physical activities

18.7 Provision of sensory gardens with various activity spaces to suit the needs of children with varied temperaments

SCORE

0 = Inadequate, 1 = Adequate, 2 = Excellent

19. Self-reflection (privacy) and small group activities	0	1	2	n/a

19.1 Social spaces where a small group of children may go to be alone (i.e., reading areas, quiet places, reflection areas, listening areas, etc.)

19.2 Space and furniture in classroom and other learning areas that provide 2–3 spaces for children to feel a sense of privacy and to control their interaction with others.

19.3 Inviting yet supervised cave spaces where students can take a deep breath, albeit momentarily, from their hectic lives.

19.4 Classrooms have clear breakout zones or breakout rooms attached to them. Breakout zones within corridors and separate from the classroom are avoided.

SCORE

0 = Inadequate, 1 = Adequate, 2 = Excellent

4 CHAPTER SALUTOGENIC DESIGN

20. Use of appropriate materials and textures	0	1	2	n/a
20.1 Use of natural materials and visible details				
20.2 Multisensory materials are used that impart qualities like smoothness, roughness, brightness, opacity, transparency, etc. to the surfaces				
20.3 Use of materials that exude warmth				
20.4 Use of glass to connect inside to outside yet making children feel secure				
20.5 Overuse of cold and hard materials is avoided				
20.6 Use of curvilinear shapes wherever possible				
SCORE				

0 = Inadequate, 1 = Adequate, 2 = Excellent

21. Homelike environments with familiar elements	0	1	2	n/a
21.1 Soft furniture, such as a couch or large armchair				
21.2 Nontoxic indoor plants are used				
21.3 Soft and comforting elements like pillows, plants, soft furnishings to add warmth and security of being home				
21.4 Other decorative touches, such as area rugs or repurposed furniture				
21.5 Provision to hang children's artwork and their pictures on the walls				
21.6 Pastel paint colors with less stimulating displays (not visually overwhelming)				
SCORE				

0 = Inadequate, 1 = Adequate, 2 = Excellent

22. Agentic learning and green/sustainable architecture	0	1	2	n/a
22.1 Spaces to learn from natural processes like sun orientation, wind flow patterns, etc.				
22.2 Visible energy conservation/sustainable measures like rainwater harvesting, solar panels, etc.				
SCORE				

0 = Inadequate, 1 = Adequate, 2 = Excellent

4 CHAPTER SALUTOGENIC DESIGN

23. Biophilic design to counter nature deficit & attention restoration	0	1	2	n/a

23.1 Ample availability of green and natural spaces in the school campus

23.2 Views of nature from inside of the classroom

23.3 Possibility of going out in the natural environment during breaks

23.4 Use of biomorphic patterns in the interior environment

23.5 Views of parking lots, roads, etc. area are avoided

23.6 Restorative spaces with items such as soft furnishings, plants, animals, window seat or aquarium are generously available

SCORE

0 = Inadequate, 1 = Adequate, 2 = Excellent

24. School pets and empathy	0	1	2	n/a

24.1 Outdoor spaces conducive for pets along with safety concerns of children

24.2 Presence of farm area for children to tend to

SCORE

0 = Inadequate, 1 = Adequate, 2 = Excellent

25. Student participation in planning and design	0	1	2	n/a

25.1 Students maintain their own green patch

25.2 Students volunteer for renovations and refurbishments in school

25.3 Student participation was considered during design and planning process

SCORE

0 = Inadequate, 1 = Adequate, 2 = Excellent

26. Stimulating playgrounds and sense of adventure	0	1	2	n/a

26.1 Opportunities for tree climbing and innovative play with movable parts

26.2 Ample space for running, jumping and other age-appropriate activities

26.3 Presence of safety nets and other safety measures to avoid injury

26.4 Proximity of school infirmary from play areas

SCORE

0 = Inadequate, 1 = Adequate, 2 = Excellent

4 CHAPTER SALUTOGENIC DESIGN

27. School configuration and smaller learning communities	0	1	2	n/a
27.1 The school is split into small learning communities through blocks or levels				
27.2 Each learning community has its own open spaces and other indoor areas for socialisation				
27.3 Each community has sufficient transparency to allow constant passive supervision				
SCORE				

0 = Inadequate, 1 = Adequate, 2 = Excellent

TOTAL SCORE	
PERCENTAGE SCORE	

SCORE RULES	
86 % - 100 %	Excellent
71 % - 85 %	Very Good
51 % - 70 %	Acceptable
31 % - 50 %	Unacceptable — Needs Work
0 % - 30 %	Poor — Needs Substantial Changes

Score rules and three point rating scale adapted from: The classroom rating scale in Lorraine Maxwell, "Competency in Child Care Settings: The Role of the Physical Environment," Environment and Behavior 20, no. 10 (2006); and the EDA SPACE app by Education Design International

The COVID-19 pandemic underscores the need to place a high priority on the mental and physical health and well-being of children in the school setting.

Conclusion

The COVID-19 pandemic underscores the need to place a high priority on the mental and physical health and well-being of children in the school setting. A salutogenic design approach to the planning of school environments aims to go beyond the traditional architectural considerations of aesthetics and academics to the more subtle but equally critical goals of achieving a built environment that promotes good holistic health: lowering stress and anxiety in children, reinforcing their sense of self-worth, and promoting individual self-actualisation—all while creating a vibrant educational community that meets the needs of mind, body, and spirit. The design guidelines and the assessment checklist as outlined above call for resources to be directed towards the construction of salutogenic learning environments for children as they prepare for the daunting challenges of the 21st century.

A Pictorial Essay

As evidenced from the HSAT (Healthy School Assessment Checklist) detailed in this White Paper, a comprehensive approach is needed to ensure that learning environments support children's health and well-being.

In this section of the White Paper, we have included a number of images that illustrate various ways in which healthy environments can be created for children of all ages. The purpose of these images is not to provide specific design solutions as much as it is to demonstrate qualities like aesthetics, good daylighting, the proper use of color, age-appropriate furnishings, personal space, comfort, connections to nature and so on.

It is not enough to create well-designed spaces for learning. It is also important to be intentional about its qualities that will directly contribute to student health and well-being.

4 CHAPTER SALUTOGENIC DESIGN

4 CHAPTER SALUTOGENIC DESIGN

4 CHAPTER SALUTOGENIC DESIGN

4 CHAPTER SALUTOGENIC DESIGN

4 CHAPTER SALUTOGENIC DESIGN

4 CHAPTER SALUTOGENIC DESIGN

4 CHAPTER SALUTOGENIC DESIGN

4 CHAPTER SALUTOGENIC DESIGN

4 CHAPTER SALUTOGENIC DESIGN

4 CHAPTER SALUTOGENIC DESIGN

115

4 CHAPTER SALUTOGENIC DESIGN

4 CHAPTER SALUTOGENIC DESIGN

117

4 CHAPTER SALUTOGENIC DESIGN

4 CHAPTER SALUTOGENIC DESIGN

119

4 CHAPTER SALUTOGENIC DESIGN

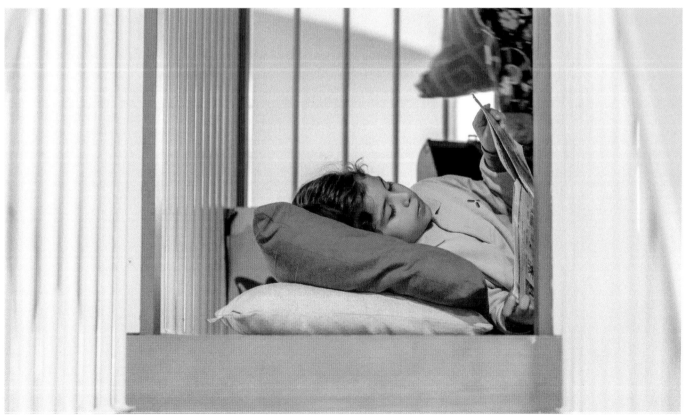

References

Antonovsky, A. (1979). Health, stress and coping. Jossey-Bass.

Antonovsky, A. (1987). Unraveling the mystery of health. How people manage stress and stay well. San Francisco: Jossey-Bass.

Arthur, P.L., & Passini, R. (1992). Wayfinding: people, signs, and architecture. Focus Strategic Communications.

Barrett, P. S., Zhang, Y., Davies, F., & Barrett, L. (2015). Clever classrooms: Summary report of the HEAD project. University of Salford. https://core.ac.uk/download/pdf/42587797.pdf

Boorse, C. (1997) A rebuttal on health. In J. M. Humber., R. F. Almeder (Eds.) What Is Disease? Biomedical Ethics Reviews. Humana Press, https://doi.org/10.1007/978-1-59259-451-1_1

Brüssow, H. (2013). What is health? Microbial Biotechnology, 6(4), 341–348, https://doi.org/10.1111/1751-7915.12063
Bureau of Indian Standards. (2005). National Building Code of India. https://law.resource.org/pub/in/bis/S03/is.sp.7.2005.pdf

Cardellino, P., Leiringer, R.& Croome, D. C. (2009). Exploring the role of design quality in the building schools for the future programme. Architectural Engineering and Design Management, 5(4), 249–262, https://doi.org/10.3763/aedm.2008.0086

Day, C. (2007). Environment and children: Passive lessons from the everyday environment. Elsevier Architectural Press.

Deci, E. L., & Ryan, R. M. (1985). The general causality orientations scale: Self-determination in personality. Journal of research in personality, 19(2), 109-134. https://doi.org/10.1016/0092-6566(85)90023-6

Dilani, A. (2008). Psychosocially supportive design: A salutogenic approach to the design of the physical environment. Design and Health Scientific Review, 1(2), 47–55. https://www.researchgate.net/publication/265349464_Psychosocially_Supportive_Design_A_Salutogenic_Approach_to_the_Design_of_the_Physical_Environment

Dilani, A. (2015, June). The beneficial health outcomes of salutogenic design. Design & Health Scientific Review, 18–32 https://dilani.org/Alan_Dilani_WHD_June_2015.pdf

Erwine, B. (2016). Creating sensory spaces: The architecture of the invisible. Routledge Taylor & Francis Group.

Essa, S. S. (2020). The impact of salutogenic factors on the process of patient's recovery case study; Erbil city hospitals. Anbar Journal of Engineering Sciences, 8(3), 228–251. https://www.iasj.net/iasj/article/183499

Evans, G. W., & McCoy, J. M. (1998). When buildings don't work: The role of architecture in human health. Journal of environmental psychology, 18(1), 85–94. https://doi.org/10.1006/jevp.1998.0089

Golembiewski, J. (2012). Salutogenic design: The neurological basis of health-promoting environments. World Health Design: Architecture, Culture, Technology, 5(3), 62–69. https://www.semanticscholar.org/paper/Salutogenic-Design%3A-The-neurological-basis-of-Golembiewski/3d5f8bc94c82da868fdde8172350efda4c69757c

Gupta. Shivam S., Gupta. Satyam S., (2019, September 15). A review on Holistic concept of health along with the correlation of salutogenesis and Ayurvedic system of medicine. International Journal of Research in Indian Medicine, 3(4). http://www.ayurline.in/index.php/ayurline/article/view/243

Hembree A., & Sholder E. (2013) Engaging holistic health through interactive design in public space—Part 1. [Architecture senior theses, Syracuse University]. https://surface.syr.edu/architecture_theses/229/

Hughes H., Wills. J, & Franz. J. (2019). School spaces for student wellbeing and learning. Springer Nature.

Kaplan, R., & Kaplan, S. (1989). The experience of nature: A psychological perspective. Cambridge University Press.

Khare, R., Mullick. A., & Raheja. G. (2011). Universal design India principles—A collaborative process of developing design principles. Slide ToDoc. https://slidetodoc.com/universal-design-india-principles-2011-a-collaborative-process/

Krause, C. (2011). Developing sense of coherence in educational contexts: Making progress in promoting mental health in children. International Review of Psychiatry, 23(6), 525–532, https://doi.org/10.3109/09540261.2011.637907

Lad, V. D. (2002). Textbook of Ayurveda : Fundamental principles of Ayurveda, Volume One, The Ayurvedic Press.

Lusher, R. H. & Mace, R. L. (1989). Design for physical and mental disabilities. In J. Wilkes & R. Packard (Eds.), Encyclopedia of architecture: Design engineering and construction/Volume 3.Wiley.

Ma, X. (2003). Sense of belonging to school: Can schools make a difference? The Journal of Educational Research, 96(6), 340–349. https://doi.org/10.1080/00220670309596617

Maslow, A. H. (1943). A theory of human motivation. Psychological Review, 50(4), 370–96. https://psycnet.apa.org/doi/10.1037/h0054346

Maslow, A. H. (1998). Toward a psychology of being. John Wiley & Sons, Inc., pp. 104–111, 123–125.

Morandi, A., Tosto, C., Di Sarsina, P. R., & Dalla Libera, D. (2011). Salutogenesis and Ayurveda: Indications for public health management. EPMA Journal, 2(4), 459–465. https://dx.doi.org/10.1007%2Fs13167-011-0132-8

Myerson, J. & Lee, Y. (2010) 'Inclusive design research initiatives at the Royal College of Art'. https://researchonline.rca.ac.uk/416/

Nair, P., Fielding, R., & Lackney. J. A. (2009). The language of school design: Design patterns for 21st century schools. Education Design Architects.

Nair, P (2014). Blueprint for Tomorrow. Redesigning Schools for Student-Centered Learning. Harvard Education Press

Nair, P., Zimmer Doctori, R. and Elmore, R. (2019). Learning by Design. Live | Play | Engage | Create.

Nordenfelt, L. (2006). The concepts of health and illness revisited. Medicine, Health Care and Philosophy, 10 (1), 5–10. https://doi.org/10.1007/s11019-006-9017-3

Osmon, F. L. (1971). Patterns for designing children's centers. Educational Facilities Laboratories, pp. 43–45.

Sanoff, H. (2001). School buildings assessment methods. (ED448588). ERIC. https://eric.ed.gov/?id=ED448588

Sharma, H., Chandola, H. M., Singh, G., & Basisht, G. (2007). Utilization of Ayurveda in health care: an approach for prevention, health promotion, and treatment of disease. Part 1—Ayurveda, the science of life. The Journal of Alternative and Complementary Medicine, 13(9), 1011–1020. http://doi: 10.1089/acm.2007.7017-A

Tanner, C. (2000). The influence of school architecture on academic achievement. Journal of Educational Administration. 38. 309–330. https://doi:10.1108/09578230010373598

Tanner, C. (2009). Effects of school design on student outcomes. Journal of Educational Administration, 47(3). https://doi:10.1108/09578230910955809

Thurber, C. A., & Malinowski, J. C. (1999). Environmental correlates of negative emotions in children. Environment and Behaviour, 31(4), 487–513.

Walden, R. (2015). The school of the future: conditions and processes–Contributions of architectural psychology. In Schools for the Future (pp. 89–148). Springer, Wiesbaden. https://doi.org/10.1007/978-3-658-09405-8_5.

Weinstein, C. S., & David, T. G. (1987). Spaces for children: The built environment and child development. Plenum Press. World Health Organization. (1999). Guidelines for Community Noise. http://www.who.int/docstore/peh/noise/guidelines2.html.

World Health Organization. (2021). The 1st international conference on health promotion, Ottawa, 1986. https://www.who.int/teams/health-promotion/enhanced-wellbeing/first-global-conference

4 CHAPTER SALUTOGENIC DESIGN

Prakash Nair, AIA
Roni Zimmer Doctori
Gary Stager
Anna Harrison

OUTDOOR LEARNING
Leave the Classroom Behind

Introduction

As of this writing, almost all 76 million school-age children in the United States are locked out of school with prospects of returning in a fulltime capacity anytime soon dimming by the day. There is a common misconception that, as one of COVID-19's worst-hit countries, the U.S. is holding back children from returning even as other countries have reopened their schools. The reality is that almost a billion students in 143 countries will not be returning to school this fall.

With COVID-19 seemingly here to stay, and pressure mounting for schools to reopen, there is now a serious move to take learning outside, as a way to maximize the number of students who can be in school at the same time. Social distancing is easier outside where space is not as much of a concern as it is within the confines of a classroom. It is likely that outdoor learning, until very recently a novelty, will soon become quite widespread.

There are many reasons behind the push to reopen schools and bring more children back. These must be balanced against the risk of propagating the virus as millions of students return to class. For example, New York City has decided that schools will not reopen if Coronavirus infection rates exceed 3%. Assuming that such safety parameters for community spread of the virus are met, and when schools do reopen, they can still do so only partially. Most schools will not be able to bring all students back on opening day because of the need for social distancing in the classroom and common areas occupied by students. Thus it is logical to ask how the problem of inadequate space might be mitigated by using outdoor areas for learning.

The reality is that almost a billion students in 143 countries will not be returning to school this fall.

5 CHAPTER OUTDOOR LEARNING

Outdoor Learning is Not New

FIGURE 62. A "classroom on a ferry". This is how New York City responded to the tuberculosis pandemic in the early 1900's. This model of outdoor learning actually "worked" insofar as none of the children got sick. This model of simply taking indoor furniture outside so that the teacher could continue to hold court to a passive and captive student audience is no longer applicable in today's world and, yet, that is what many schools are now considering due to COVID-19 – to use the outdoors to do exactly what they had been doing indoors. This paper argues strongly against such an approach.

Bureau of Charities, via Library of Congress

The subsequent New England winter was especially unforgiving, but children stayed warm in wearable blankets known as 'Eskimo sitting bags' and with heated soapstones placed at their feet. The experiment was a success by nearly every measure — none of the children got sick. Within two years there were 65 open-air schools around the country either set up along the lines of the Providence model or simply held outside. In New York, the private school Horace Mann conducted classes on the roof; another school in the city took shape on an abandoned ferry."

With "open-air" schools proliferating everywhere, the US Department of the Interior put out a bulletin on the subject in 1916.

"Forest Schools" fully embrace the idea of outdoor learning for younger children.

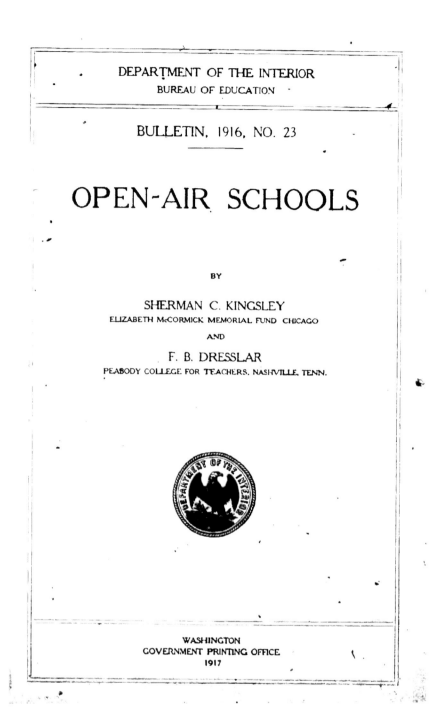

FIGURE 63. The previous Open-Air Schools initiative was backed by the US Department of the Interior. This 280-page manual was written as a guide for schools who wanted to take learning outdoors. At that time, the focus was still on "outdoor classrooms" which is also something this paper recommends against.

Since then, while outdoor learning has been celebrated at many schools, it has not caught on in a big way. Today, only a few schools have the outdoors as the central tenet of their learning philosophy. "Forest Schools" fully embrace the idea of outdoor learning for younger children. Rain or Shine Mamma is a group that promotes forest schools with two great books on the subject. They note, "Researchers studying forest schools have found that outdoors, children hone their motor skills, engage in more creative play, have fewer conflicts, stay healthier, learn to be more independent and develop a compassion for nature and wildlife that is likely to last a lifetime.

The first U.S. forest school, Cedarsong Nature School, started in 2006. Today nature-based preschools in this country number in the hundreds. This growth is encouraging to all of us who care about raising children who are connected with nature!"

Beyond school, the outdoors is a booming business. An entire industry (summer camps) has been built to take advantage of outdoor activities and, because of that, schools may not face as much resistance from parents to outdoor learning as they may have in the past.

5 CHAPTER OUTDOOR LEARNING

Don't Simply Take the Classroom Outside

In spite of overwhelming evidence that being outdoors is good for children, outdoor areas in schools remain woefully underutilized. Even in the most crowded schools, during most of the school day, the outdoors are windswept deserts with hardly a student in sight. For ease of maintenance, much of the outdoor areas are paved, especially in urban areas most in need of green spaces.

It is tempting to see all the unused outdoor areas simply as an answer to the space crunch that schools are experiencing and use them to quickly provide more teaching areas. In other words, create more "outdoor classrooms." This bland approach that looks single-mindedly at the logistical challenge of "housing" more students is one that has its advocates, because it presents what is seemingly a low-hanging-fruit option.

A Fast Company article that makes a strong argument for outdoor learning falls prey to a distressing, but seemingly universal, mindset that real learning cannot happen unless students are sorted by age and confined to a classroom under the watchful eye of one adult who then orchestrates their every move. The image (Figure Three) they chose to highlight outdoor learning could just as easily have been used to discuss indoor schooling in the 1950s.

The prevalence of the mistaken belief about bringing classrooms outside is reinforced in this article in the New York Daily News titled, How to create outdoor classrooms: Maximize use of schoolyards to get in-person teaching to as many students as possible. This piece was written by the City's comptroller in what was undoubtedly a well-meaning but misguided attempt to commandeer outdoor spaces to increase teaching – as opposed to learning. The article notes that New York City has 29.5 million square feet of outdoor yard and physical education spaces at its 1,575 schools, "which would allow students to stay outside but still maintain easy access to bathrooms, handwashing stations and cafeterias." It ends with this call for action, "Education is the foundation of a healthy, equitable, and thriving city. Let's meet this moment with boldness and creativity." Unfortunately, there is nothing bold or creative about throwing kids outside into makeshift classrooms so that they can be taught the same boring content in spaces

FIGURE 64. This is the photo that Fast Company chose as its marquee image to showcase national efforts to take learning outdoors. It reinforces a shortsighted idea that outdoor learning is simply a stopgap measure in which classrooms are, literally, moved outdoors until children can be safely brought indoors again after COVID-19.
Photo: Morning Brew/Unsplash

5 CHAPTER OUTDOOR LEARNING

FIGURE 65. *Contrast this photo with Figure 64. What's good about it is the environment. Children have nature views, shade, and are breathing fresh air. A drum circle like this one in Florida teaches a skill while keep students active and engaged. This is much preferred to just taking a "class" outside so students can be lectured to.*

that are likely to be even more uncomfortable than the prison-like boxes within the school building itself. This may sound like a harsh assessment, but a rushed solution that could adversely affect the life of over one million children should not be sugar-coated. While this approach of grabbing whatever space is available to bring back more students may have had its place in 1907 during the height of the tuberculosis pandemic, we do have better options today.

So what's wrong with classrooms? That question is answered very simply – classrooms are obsolete. I wrote about this in Education Week a few years ago in which I noted, "The classroom is a relic, left over from the Industrial Revolution, which required a large workforce with very basic skills. Classroom-based education lags far behind when measured against its ability to deliver the creative and agile workforce that the 21st century demands."

 .. there is nothing bold or creative about throwing kids outside into makeshift classrooms.

5 CHAPTER OUTDOOR LEARNING

Outdoor Spaces are Therapeutic

Contrast the New York City proposal with what Green Schoolyards America is advocating. Their recent whitepaper on the subject of outdoor learning is titled, A Proposal to Engage School Grounds and Parks as Strategic, Cost-Effective Tools for Improving Academic, Mental and Physical Wellbeing as Schools Reopen. According to Green Schoolyards, "Outdoor Spaces are Therapeutic. Students and staff may return to school with stress and trauma associated with isolation, uncertainty, and illness. To ease the burden, nature-rich outdoor areas identified and developed on or near each campus can provide quiet, reflective spaces to unwind and relax." Green Schoolyards is not opposed to the idea of "Outdoor Classrooms" but they would do it in a "nature-rich" environment, and that makes all the difference. Theirs is an approach of housing students outside as part of a holistic solution to student wellbeing as opposed to those who are simply looking for extra space so more students can attend school.

The Green Schoolyards white paper lists several reasons why the use of outdoor learning can mitigate the negative impacts of COVID-19 on:

1. Equity: Not everyone has benefited in the same way from online learning. The children most in need of help tended to be the ones who got it the least, and this trend will continue without serious interventions

2. Learning: Even for those students who successfully made the switch to the online model, disrupted schedules, distractions and the unavailability of face-to-face coaching have adversely affected their learning

3. Physical Health: With most of their time spent in front of a computer screen, student stuck at home have had far fewer opportunities for physical fitness than they would have had at school

4. Mental Health: As with the equity issue above, students who most need the structure and caring provided by the adults and their friends at school miss out the most when school is not in session

5. Economic Health: Many parents are unable to work and take care of young children who would otherwise have been in school. This is a significant segment of the population whose financial struggles carry over into the economy at large

6. Education Workforce: The education sector employs many tens of thousands of teachers, administrators and staff. Even though a majority of teachers may be able to retain their jobs until schools reopen, many others not considered "essential" will lose their jobs and contribute to the overall decline of the national economy.

FIGURE 66. At most schools, such areas between buildings are left barren and untended. Here, at Corbett Prep School in Tampa, every available part of the outdoors is used to connect children with nature and encourage outdoor activities. It is a strategy that would also work well on tight urban sites.

5 CHAPTER OUTDOOR LEARNING

FIGURE 67. This treehouse at Anne Frank Inspire Academy in San Antonio, Texas, by famed treehouse master Pete Nelson makes an otherwise nondescript small school site special and has the potential to create many happy memories for the students who study here.
Photo: Prakash Nair

Will Older Students Be Deprived of a Quality Education Outdoors?

There is a prevalent belief that older students cannot afford the luxury of puttering around outside when they have to be doing "serious" work inside with teachers. Before we look at how the serious work that high schoolers need to do may be better accomplished outside, let us dispel the myth that they are getting a lot of benefit from their indoor classroom setting.

"Classrooms are based on the erroneous assumption that efficient delivery of content is the same as effective learning. Environmental scientists have published dozens of studies that show a close correlation between human productivity and space design. This research clearly demonstrates that students and teachers do better when they have variety, flexibility, and comfort in their environment—the very qualities that classrooms lack." Content knowledge that is disproportionately emphasized during high school can be more easily acquired online. In the online world, high-schoolers can learn at their own pace from their own classroom teacher or choose from among hundreds of content specialists, colleges and universities who offer free online courses. For all the in-person experiences that older students need, the outdoors provides a much better alternative than classroom learning. This includes things like research, learning with mobile technologies, team collaboration, cooperative learning, one-on-one time with the instructor, hands-on learning and environmental projects.

 Classrooms are based on the erroneous assumption that efficient delivery of content is the same as effective learning.

5 CHAPTER OUTDOOR LEARNING

Outdoor Learning Fights Nature Deficit Disorder

In an influential New York Times Opinion piece, Timothy Egan credits Richard Louv with coining the term "Nature Deficit Disorder" and cites Louv's 2005 book, Last Child in the Woods, which says that "kids who do play outside are less likely to get sick, to be stressed or become aggressive, and are more adaptable to life's unpredictable turns."

My recommendation that learning spaces should be connected visually and physically to nature is supported by Kaplan and Kaplan's Attention Restoration Theory. This theory provides a framework for identifying environments with "soft fascinations" that are restorative and beneficial for cognitive performance. Soft fascinations are scenes or objects that one can observe with effortless attention, for example, leaves rustling in the wind, water running over pebbles in a creek or clouds slowly moving in the sky.

Soft fascinations are scenes or objects that one can observe with effortless attention.

Outdoor Learning in Urban Areas

Even though children in urban areas are more likely than suburban or rural children to suffer from nature deficit disorder, urban schools tend to have fewer opportunities for sending students outdoors, in communion with nature. Many urban sites are so highly built up that they leave little room for green areas and restful outdoor zones. However, with a

FIGURE 68. Even in situations with limited space, it is possible to provide opportunities for outdoor learning immediately adjacent to the building.

5 CHAPTER OUTDOOR LEARNING

FIGURE 69. Here at St. Martin de Porres High School in Cleveland, Ohio, an urban garden is maintained by the students for the benefit of the community.

little imagination, underutilized areas can be converted at minimal expense to serve outdoor learning activities.

My advice to urban schools is to think small, as in a small vegetable patch, a small fishpond, a small seating and reading area, or a small fountain. Where there is, literally, no room on the site to locate outdoor learning activities, consider using an available verandah or balcony or the roof of the school building for learning activities. Rooftop learning may include vegetable gardens, weather stations, and canopied areas for individual and group seating for activities such as reading, research, independent study, and team collaboration.

 My advice to urban schools is to think small, as in a small vegetable patch, a small fishpond, a small seating and reading area, or a small fountain.

What Does The Research Say?

Those who vigorously demand that children return to school ignore the reality that the design of traditional school buildings goes counter to everything we know about child development, social and emotional growth and personal identity. All these needs are subsumed by the institutional goals of efficient management of children and delivery of content. The classroom is the result of this unhealthy focus on efficiency, but the opportunities to find balance by going outdoors was always there. With COVID-19 bringing outdoor learning to the forefront of discussions regarding school reopening, questions will be asked about its efficacy and whether we are shortchanging children's learning by taking them out of the classroom.

Keeping aside the benefits of bringing children safely back to school already discussed in this paper, let us look at what the research is saying about outdoor learning. Overwhelmingly, the evidence supports the incorporation of outdoor learning as an integral part of every student's school day. Here is what we know:

1. Children with symptoms of ADHD are better able to concentrate after contact with nature
2. Play in a diverse natural environment reduces or eliminates bullying
3. Nature helps children develop powers of observation and creativity and instills a sense of peace and being at one with the world
4. Early experiences with the natural world have been positively linked with the development of imagination and the sense of wonder ,
5. Wonder is an important motivator for lifelong learning
6. Children who play in nature have more positive feelings about each other
7. A decrease in children's time spent outdoors is contributing to an increase in myopia in developed countries
8. Outdoor environments are important to children's development of independence and autonomy

> *Overwhelmingly, the evidence supports the incorporation of outdoor learning as an integral part of every student's school day.*

FIGURE 70. *This is a generously sized vegetable garden at Learning Gate Community School in Lutz, Florida, where students do most of the planting and tending. A substantial part of every student's day at this school is spent outdoors and this was true even before the COVID-19 pandemic.*

Does Outdoor Learning Support Proven Educational Strategies?

There are a few educational strategies that many of the most progressive schools and school districts in the U.S. and around the world have endorsed in principle and, to a lesser extent, in practice. It should come as no surprise to anyone that gains in these areas have been hard fought. Consequently, a fair question that should be asked of any change being forced upon educators by COVID-19 is if it will endanger the progress that has been made to date. In this paper, I will show that outdoor learning poses no threat to educational progress. In fact, should it emerge as a central theme in education, outdoor learning would have the potential to enhance, accelerate and more equitably distribute the benefits of a relevant, well-rounded education to all students.

The list below includes five key strategies championed by progressive educators everywhere in public, private and parochial schools and discusses how they are all enhanced when children are permitted to learn outside.

Student-Centered Learning: The four elements of student-centered learning include 1) Personalization: It is obviously easier to personalize learning in an environment where students don't have to all be doing the same thing at the same time – easier outside than in a classroom; 2) Individual Mastery: Whereas in the classroom, students have fewer ways in which to demonstrate their understanding of and aptitude for a subject, the outdoors provides a richer palette of learning modalities to enable students to demonstrate their in-depth grasp of a subject or topic; 3) Anytime Anywhere Learning: The traditional school building puts mobile technology into the hands of stationary students. With more space to move around, students will use technology when, where and as needed to make learning truly anytime anywhere and; 4) Student Ownership and Agency: The space limitations imposed by classrooms force teachers to standardize lessons in ways that leave little room for individual preferences. Outside, students have room to spread out and do different things. Having outdoor space available for learning means a broader scope and inspiration for individual students' customized short-term and long-term projects.

Technology Integration: The use of technology can be supported in outdoor spaces by providing subject matter for photography and video narrative that is very limited in a classroom setting. With the increasing range and bandwidth offered by wireless networks along with reduced costs, shaded outdoor areas adjacent to indoor areas can be used to work with mobile technologies like laptops, tablets and smart phones. Outdoor learning also exposes students to technologies they would not be able to use within a classroom such as survey instruments, water and soil testing equipment, and even some older "technologies" like sundials and weather vanes.

FIGURE 71. *If an outdoor classroom is essential, then a green amphitheater like this one at Swarthmore College is an elegant solution, since it can permit social distancing and allow for many more modes of learning, such as music, dance and performance, student presentations and quiet reading. It is also a healthy, comfortable and inspiring space with lots of fresh air. Compare this to a traditional indoor classroom!*
Photo: Swarthmore College

Flexible Scheduling: Not being at the mercy of a fixed schedule means that a student who is passionate about a particular project is not always forced to interrupt his or her work because the bell has rung at the end of the 50-minute period. Outdoor settings are much better for hands-on projects that are not as time-bound as fixed subjects offered at fixed times. Flexible scheduling that is based on students moving on only after having achieved mastery as opposed to simply having served time in class is more practical in hands-on settings like the outdoors.

Teacher Collaboration: The creative use of a permeable indoor-outdoor space can allow two teachers to work together, for instance where one teacher is working outside with a small group of students, conferencing or providing seminar-style instruction, while the other supervises students at work indoors. Beyond that, with hands-on learning more easily done outdoors, two or more teachers can create interdisciplinary projects and work collaboratively throughout the semester.

A Positive School Climate: Functional and beautiful outdoor spaces can help foster a positive school climate because they help students relax and reflect. Robin C. Moore's 1996 study revealed that children who play in nature have more positive feelings about each other. So re-vegetating and enhancing the campus's natural features can be important for social and emotional development.

Connection to the Environment: Today, more than ever, connecting to the environment is an important educational goal. This goal is most strongly supported by functional outdoor learning spaces. Connecting indoor spaces to the outdoors is just one way to make children more aware of their natural environment. Preserving and enhancing the campus's natural features including topography, watercourses, trees and shrubs, is another. These are only meaningful to students if they have a hand in the upkeep and maintenance of the school's outside environment. This can be effectively done by incorporating environmental studies into the curriculum as opposed to being presented as an after-school, extracurricular activity.

Space limitations imposed by classrooms force teachers to standardize lessons in ways that leave little room for individual preferences.

Designing Outdoor Play Areas

Outdoor play has been one of the most unfortunate casualties of COVID-19. For many children, schools had provided their only access to outdoor play and, during the pandemic, play has been replaced by time in front of a computer, phone or television screen. We intuitively know that all human beings have an intrinsic desire to play, but why is it important?

Play is important because it's how we've evolved to learn. It is also increasingly important because of rising rates of obesity among children. A 2012 survey of children by the Heart Foundation of Australia found that the kinds of play features children want are not overly manicured, instead they're a little scruffy. "Hills to roll, run and slide down, boulders and tunnels to scramble around, branches and leaves with which to build tree houses and cubbies, water features and community art. "They also want to be challenged by their play areas sometimes, to be a little bit frightened" according to a spokeswoman for the Heart Foundation. "They want to be able to build and construct things, so we need to provide moveable parts, debris, to not blow all the fallen leaves and twigs away with a blower but rather plant trees that drop their leaves." It's quite a different picture from the ubiquitous, plastic-molded, slippery slide and monkey bars that you'll find in most playgrounds.

Play spaces should pose open-ended questions. They should prompt imaginative responses. They should foster a connection between the players and the natural environment. If you are keen to invest in improving your school's play spaces, before jumping to choose a ready-made climbing gym from a catalog, consider how much better you could build it from scratch. There is great inspiration to be found online – several bloggers collect pictures and reviews of outstanding play spaces from around the world. Employ a professional designer if you can, but more importantly have students participate in the design process, and have your professional designer work alongside the students.

 Play is important because it's how we've evolved to learn.

FIGURE 72. *Designing for Play: An unstructured area with an assortment of natural elements inspires more creative play than a structured playground does.*

How Should Areas for Fieldwork be Designed?

School design needs to support current education literature on project and experiential learning that argues for the importance of hands-on, active, lifelong learning. "Fieldwork" is a word used to describe the firsthand collection of data through sensory observation in an urban, rural or natural environment. It could be as simple as two-year-olds feeling the difference between sand, mud and snow on their fingers, or five-year-olds counting the number of trees on the campus, or by ten-year-olds filming the nesting habits of a bird. In other words, fieldwork is very important and schools need to be designed to support it.

Design for fieldwork on campus largely means maintaining and enhancing the special features of the campus's natural environment. Water courses are a wonderful asset for fieldwork and can be re-vegetated to provide a habitat for native birds and animals. Wooded areas are all too often demolished to make way for development, but they make great learning places. In fact, proponents of the (German) Waldschule (Forest School) movement specifically seek out forests in which to base outdoor classrooms. Don't be intimidated by the word "forest." On many smaller urban sites, a small grove of trees can support its own ecosystem that is worth studying.

Even without these kinds of environments on campus, a school may be able to take advantage of adjacent and nearby public parks, and sites can be chosen for new school developments that take this into account.

5 CHAPTER OUTDOOR LEARNING

In the early childhood years, outdoor fieldwork looks very much like play, as toddlers and preschoolers are given time and space to freely explore and observe, using all their senses, the qualities of the natural environment. Through the elementary years, inquiry-based study of the natural environment can begin with the children's developing understanding of the observable art and science present outdoors. School designers who understand that such activities can be a valuable piece of every child's education will be able to work closely with educators to design suitable outdoor learning environments.

Students in the Environment Club at Scott Creek Primary School in Adelaide, Australia, have developed a nature discovery trail in an adjacent forest (or bushland, as it's referred to locally), for other students as well as members of the local community. Each station on the trail explores an element of the bushland, sharing facts about the location and explaining the group's revegetation project.

Fieldwork tends to be quite active, not sedentary, and as such can be supported in all but the most hostile climates, on all but the hottest or coldest 10% of days.

FIGURE 73. *Designing for fieldwork: Woods and ponds to explore and study wildlife are of more interest to students than large stretches of asphalt or grass.*

 "Fieldwork" is a word used to describe the firsthand collection of data through sensory observation in an urban, rural or natural environment.

Make Room for The Garden

Adults lament the fact that so much of our children's time is spent in front of a screen — these include phones, tablets, computers, and TV. This has come at the price of them spending more and more time indoors. By and large, schooling is also an indoor activity, but we recommend that every effort be made to move learning activities outdoors. One natural fit for an outdoor activity is gardening. All schools should make an attempt to have a kitchen garden. Where land and weather permit, schools can also lead efforts to build and maintain community gardens.

This is an activity that students tend to enjoy. It also comes with numerous ancillary benefits such as breathing fresh air, becoming more environmentally conscious, becoming more likely to eat healthy, organic fruits and vegetables, becoming more aware of good health and nutrition, being more physically active and learning about the benefits of teamwork and community building.

Look for partnerships with local organizations to help your school start a vegetable garden. In the words of a Montreal-based organic gardening group, "Imagine your schoolyard transformed, filled with fruit trees, berries and perennial vegetables. Transform your underused space into a positive space to connect and learn about plants and nature.

Children get to learn about nature, all while learning to care for the plants and each other. They discover where food comes from, how it grows and taste the fruit of their labors. Treat their sweet tooth to the real sweets of nature, strawberries, raspberries, and blueberries. Plant a fruit tree in your school yard. Integrate the garden into your cafeteria, into your biology class, or gym class, make healthy living and eating a visible priority at your school." In a similar vein, Growing Together, an Oakland, CA-based nonprofit organization, provides support for underserved communities and schools using organic gardening as a tool for education, healthy living, and community building.

 All schools should make an attempt to have a kitchen garden.

FIGURE 74. *Collingwood College in Melbourne, Australia, pioneered the Stephanie Alexander kitchen garden project where children can plant, grow, harvest, cook and eat organic foods.*

5 CHAPTER OUTDOOR LEARNING

Taking Care of Animals

Outdoor areas may be suitable at many locations for maintaining a chicken coop or even managing a small petting zoo. There are various ways in which taking care of animals in school can be a good thing for students. It teaches them empathy, responsibility, and discipline, and the bonding with animals that children naturally enjoy.

Despite their obvious benefits and the great affinity that children tend to have with animals, programs where children get to work with or take care of animals is the exception rather than the rule. The Ballarat Grammar School Farm Program is one worthy of emulating. Here, students spend most of their 4th grade year working on an active farm. This program shows how much of the learning that we believe can only happen in a classroom is actually better delivered in nature where students are breathing fresh air, learning valuable life skills, being more active physically and taking care of animals.

There are various ways in which taking care of animals in school can be a good thing for students.

FIGURE 75. *Taking care of animals in school can be a good thing for students. It teaches them empathy, responsibility and discipline, and the bonding with animals that children naturally enjoy. Photo: Joni Mulvaney*

FIGURE 76. *Outdoor learning is an integral part of each student's school day at the International School of Dusseldorf, Germany. At this school, students go out every day regardless of the weather to participate in a wide variety of engaging learning activities. ISD subscribes to the philosophy that "there is no such thing as bad weather, only bad clothing."*
Photo: Prakash Nair

How About the Weather?

Weather is not a deal-breaker when it comes to outdoor learning. Of course, most of what is being described in this paper would be easier and more logical at times when the weather is pleasant --- which is the case for most of the school year in many parts of the United States and around the world. However, other than when there is an extreme weather emergency such as a storm or blizzard, threat of tornadoes, a deep freeze, or unusually hot temperatures, outdoor learning should not be a problem. Some adverse weather conditions can be mitigated with measures such as providing shade structures and ensuring that all children wear appropriate clothing.

 Weather is not a deal-breaker when it comes to outdoor learning.

Outdoor Structures – Short and Medium Term Solutions

The downside to some of the solutions presented here like tents, domes and inflatables is that they do not provide the same level of acoustic comfort or climate control that are possible in indoor areas. That is why schools looking for spaces to house activities that would normally be conducted indoors like quiet study, lectures, collaborative work and presentations, may be better off using existing gymnasiums, cafeterias and libraries. The temporary outdoor spaces would then be more appropriately used for the noisier activities like eating and sports.

Shade Trees: Setting up a learning activity under the shade of an existing tree is the least expensive solution for taking learning outside. If there is a dearth of trees on existing school grounds, it's never too late to plant some. Schools are likely to be around a long time and certain species of trees grow fast enough to provide good shade within a few years.

Awnings: A simple way to extend learning into areas immediately adjacent to the school building is to strategically install awnings in locations that would remain in shade during most of the school day. Awnings expand learning spaces significantly and are suitable for a wide variety of outdoor activities from art and science projects to outdoor eating.

Tents: Many schools are used to the idea of erecting tents outdoors for short-term gatherings like graduations. The advantages of tents are 1) Speed and ease of installation; 2) Relatively low cost; 3) Versatility – can be used for a variety of learning and social activities; 4) Can be spread over a generous footprint providing substantial added usable area. If tents are used, it is important that they are sufficiently open as to allow abundant daylight. Translucent materials would also work well to daylight the space without glare.

Sheds: Sheds have the same advantages as tents but they are a bit more permanent and can be heated and cooled. They are more expensive to build and also take longer to install. Sheds may also need special building permits.

Traditional Greenhouses: These are basically glass sheds with a specific purpose. Tents and sheds are more about utilizing the outdoors to extend available semi-indoor spaces. On the other hand, greenhouses could more logically be considered an active outdoor space since they allow students to be working with natural elements in a fully daylit environment even in adverse weather conditions. Greenhouses are about "bringing the outdoors in" and represent a desirable use of outdoor areas.

Geodesic Domes: Geodesic domes provide an elegant lightweight building system that allows larger footprints to be enclosed without obstructions than traditional sheds. They are also more attractive and can be used either as a greenhouse or for any other kind of outdoor learning activity.

Inflatable Structures: These structures are more suitable for active use because of their poor acoustic quality. They tend to be used in more extreme climates that require heating in the winter while keeping out rain and snow.

FIGURE 77. Here is a picture that shows why there is no such thing as bad weather only bad clothing. Children playing on a snowy day at Watson School, Canada.

How Much Money is Needed?

This paper provides a wide range of suggestions with differing budget needs for the use of outdoor learning areas. The key is to adopt the least interventionist approach to take advantage of outdoor settings while providing great places for teaching and learning.

Any solution that entails the installation of a temporary or semi-permanent structure such as a tent, shed, dome or inflatable will cost money. Costs are also impacted by the extent to which efforts are made to "condition" inside air. With COVID-19, the best solution is to use as much fresh air as possible with passive heating and cooling. It would be prohibitively expensive to install a forced air heating or cooling system that will need to have all the safeguards in place so as not to spread viruses like COVID-19. Without such technologies in place, the virus could travel via the air handling system across the entire enclosed space and vastly increase the possibilities for infecting large numbers of occupants.

The key is to adopt the least interventionist approach to take advantage of outdoor settings.

FIGURE 78. *Chess as Outdoor Theater. Integrating math, social learning, focus, and being active outside. Kneeler Design, Victoria, Australia. Photo: Silvi Glattauer*

5 CHAPTER OUTDOOR LEARNING

A New Curriculum to Take Learning Outside?

The simple answer to this question is yes! Outdoor learning expands opportunities to implement a more student-centered and hands-on curriculum. To understand, let us look at the three kinds of curriculum. They are: 1) Subject centered; 2) Learner Centered and; 3) Problem centered.

Most classroom-based schools prefer the subject-centered curriculum model because of the relative ease with which it can be delivered. The physical constraints presented by different teachers in different rooms, the lack of adequate space within classrooms and the regimented ringing of the bell makes it very difficult to implement a true learner-centered or problem-centered curriculum. Further exploration of these three curriculum types within the context of what any particular school may be doing is an important pre-condition to taking learning outside.

FIGURE 79. Designing for "inside goes outside": A Variety of outdoor seating options encourages students to read and study in the fresh air and sunshine.

FIGURE 80. Creating Outdoor learning Experiences does not need to represent a great expenditure of funds as seen by this modest renovation at Hillel Academy, Tampa, Florida, in which a simple wood trellis provides shade and a quiet place to study between two buildings.

FIGURE 81. *This image, also from Hillel Academy, shows an even less expensive option of simply installing a lightweight "sail" to bring shade to a deck adjacent to a learning commons.*

How About Professional Development?

Schools learned the hard way with online learning. A vast majority of teachers (83%) found the transition from in-person to online teaching difficult. Given the speed at which COVID-19 led to school closures it is understandable that everyone, including teachers, were caught off guard. Even if schools had the resources, there simply wasn't enough time to reexamine the curriculum and offer teachers the help they needed to move all teaching and learning online.

The same is true with outdoor learning. The thesis of this paper that the outdoors provides many rich opportunities for learning requiring a rethinking of both pedagogy and curriculum. It follows that teachers will need appropriate professional development to transition properly to a model in which students move freely between indoor and outdoor spaces.

Given the speed at which COVID-19 led to school closures, it is understandable that everyone, including teachers, were caught off-guard.

Professional development efforts should focus on how outdoor learning experiences can support a student-centered, and problem-centered approach to learning. Training should help teachers with; 1) Working as a team; 2) Tailoring work for students so that it can be personalized; 3) Creating an assessment system that measures process and progress and also soft skills like social and collaboration skills; 4) Encouraging the design and deployment of interdisciplinary projects and problems; 5) Looking at the particular characteristics of each outdoor locale that is available in order to create assignments and activities that maximize that particular location's learning potential; 6) Ensuring that students are partners in the process having a say in what, how and where they learn.

Professional development efforts should focus on how outdoor learning experiences can support a student-centered and problem-centered approach to learning.

Building Partnerships

The good news is that schools don't need to do all this on their own. This is a perfect time to partner with available public, private and community organizations who may have good experiences with outdoor learning.

Larger towns and cities will have more opportunities to find local partners, but even small towns can benefit from regional and national initiatives. A great example of this is the National COVID-19 Outdoor Learning Initiative led by Green Schoolyard America in partnership "with three other organizations in the San Francisco Bay Area, the San Mateo County Office of Education, the Lawrence Hall of Science museum in Berkeley, and the environmental education nonprofit Ten Strands."

The good news is that schools don't need to do all this on their own.

Outdoor Learning—During and After COVID-19

As of this writing, the pressing issue for schools is logistical—how to bring children and teachers back safely in the midst of a pandemic. That is the primary lens through which outdoor learning opportunities is likely to be viewed.

Social Distancing Hard to Do in Schools: Enforcing social distancing, particularly for the younger grades, will be very difficult for teachers. No teacher wants to be cast in the role of a "warden" constantly monitoring children and making sure they do not get close to each other. From a

Enforcing social distancing, particularly for the younger grades, will be very difficult for teachers.

practical standpoint, this is nearly impossible to sustain for any length of time. Yes, the outdoors provides more opportunities to maintain social distance but, even here, teachers will have to severely limit what children can do. The end result will be more overt teaching and less student participation.

This problem can be addressed (both indoors and outdoors) by selecting the "bubble" or cohort option where groups of students are kept together at all times and do not mingle with other cohorts. With the cohort system, different groups of students can occupy different sections of the outdoor learning areas so that they do not come in contact with each other. Under this system, social distancing would be unnecessary and school as we know it can continue unimpeded. Cohort-based groupings actually make the imperative for outdoor learning even more urgent, because, otherwise the temptation would be to keep groups of students trapped in their classrooms all day so as to prevent them from coming into contact with students from other groups. This would put enormous strain on both teachers and students and won't be tenable in the long run either.

At the start of this paper we saw how this kind a single-minded focus on logistics as opposed to what is good for children will result in solutions that provide, at best, a band-aid fix as "classrooms", literally, move outside as was the case with the previous pandemic in the early 1900s.

We can do better. Yes, it may take a little longer to come up with the right solutions but the learning benefits will be substantial. More important, changes to the way in which our children are educated that incorporate rich outdoor learning experiences can and should continue beyond COVID-19. To cash-starved communities who are grappling with the challenges of growing enrollment and inadequate space, outdoor learning can provide much-needed extra space at very low cost. The good news is that, far from being a compromise, this approach will actually enhance learning at all levels. That is why every school and school district, regardless of its financial resources, should make outdoor learning an integral part of its curriculum.

Cohort-based groupings actually make the imperative for outdoor learning even more urgent.

Where to Start?

A good place to start would be to distribute copies of this paper to the wider school community to build support for outdoor learning. The various resources listed at the end of this paper can help build the case to take learning outside.

With everything else they have to worry about, educators shouldn't have to figure out exactly how to make this happen. That is why the process should be led by a professional architect or landscape architect with impeccable educational credentials. They should be teamed up with an internal Outdoor Learning Leadership Team (no more than 10 to 12 people) including representatives from educational leaders, Board members, teachers, students, parents, and benefactors from local business and community organizations.

The process itself will have these essential steps:

Discovery: Through a series of remote meetings and workshops and at least one or two in-person visits, the following information will be collected: a) What is the current situation with regard to plans for reopening school(s); b) what are the available resources and opportunities for outdoor learning; c) what are the community's aspirations and educational priorities; and d) What is the available budget and how much can be raised in the short, medium and longer (two-to-five-year term)?

Open Space Efficacy Assessment: Today, there are several sophisticated space assessment tools that can be used to provide a quick summary regarding the potential of each school's outdoor learning opportunities. These

With everything else they have to worry about, educators shouldn't have to figure out exactly how to make this happen.

tools can be fully customized to meet each school's specific situation. The purpose of the assessment is to create a "gap analysis" between what exists today and what it will take to achieve the educational institution's educational goals.

Assessments will provide both a qualitative review of the outdoor spaces that are available as well as a numeric "score" between 0% to 100% for each school to indicate the current quality of its outdoor learning potential. Naturally, most schools that have not already incorporated outdoor learning as an integral part of its curriculum will score low on this measure. The good news is that as efforts get underway to take advantage of the available opportunities – some without the expenditure of any funds at all – the outdoor learning score will go up. Naturally, as the school begins to implement its strategic plan (below), it will start to see dramatic improvements in the outdoor learning assessment score – thus providing a concrete measure and reassurance to all stakeholders about the success of their efforts.

Strategic Plan: This will vary based on whether or not the effort is for a single school or a district-wide initiative. The strategic plan will flow naturally from the results of the Discovery and Assessment steps noted above. It will provide all the information needed to implement the decisions made by the educational leadership team.

School-Based Master Plans: This is the "design" of the outdoor learning initiative. It will include two parts – the physical elements that will be addressed at each school and the educational components such as changes to the curriculum and professional development plan. The Master Plan will also include a schedule for the changes and a phasing plan assuming that some of the more expensive elements of the plan may take longer to build and deliver.

Implementation: This step includes the actual work needed to make selected outdoor areas available for use as planned. The extent of work needed including any temporary structures and landscaping will depend upon the decisions memorialized in the strategic plan.

Post Occupancy Assessment: The same assessment tools used to benchmark the school site at the start of the process will be used periodically to measure progress as each phase of the project is completed. This measure will serve as a formative tool to continuously improve teaching and learning in the selected outdoor areas.

This can be the real and lasting legacy that future generations of children living healthy, balanced lives and fully enjoying the outdoors will remember.

Conclusion

COVID-19 has been a nightmare for entire societies that are in disarray and schools have been among the hardest hit insofar as they are, rightfully, more cautious about reopening. At first, there was hope that it would all be over by the summer of 2020 and things, including education, could return to normal by the fall. Indeed, a few countries have returned to some semblance of life before the pandemic. In the United States, the day when full victory from the pandemic can be declared may be many months or even a year or more away. Even after a full return to normalcy, the social and emotional scars are likely to take a long time to heal fully. That is all the more reason why outdoor learning initiatives such as those described in this paper are important. This can be the real and lasting legacy that future generations of children living healthy, balanced lives and fully enjoying the outdoors will remember when they look back at the COVID-19 pandemic of 2020.

Endnotes

[1] Ryan Heath. Open schools are the exception, not the rule, around the world. American children are among more than a billion students globally facing a fall without traditional school. Many may never return to the classroom. Politico, July 30, 2020 https://www.politico.com/news/2020/07/30/open-schools-exception-not-rule-387507

[2] Gina Bellafante. Schools Beat Earlier Plagues with Outdoor Classes. We Should Too, New York Times, July 17th 2020.

[3] Sherman C. Kingsly and F.B. Dresslar. Open Air Schools. Bulletin 1916, NO. 23, Department of the Interior, Bureau of Education
https://www.dropbox.com/s/ajrcsve8z7aqyja/Open%20Air%20Schools%201916%20Dept%20of%20Interior.pdf?dl=0

[4] The two outdoor learning books promoted by Rain or Shine Mamma are: 1) Akeson McHurk. There's No Such Thing as Bad Weather; and 2) Classroom with No Walls. The Power of Outdoor Learning. For more information on these and other resources please visit: http://rainorshinemamma.com/what-is-forest-school/

[5] What is Forest School by Rain or Shine Mamma http://rainorshinemamma.com/what-is-forest-school/

[6] Inside the quest to reopen schools – by moving classes outside by Nate Berg, Fast Company, July 20, 2020 https://www.fastcompany.com/90532401/inside-the-quest-to-reopen-schools-by-moving-classes-outside

[7] Scott Stringer. How to create outdoor classrooms: Maximize use of schoolyards to get in-person teaching to as many students as possible, New York Daily News, July 29, 2020 https://www.nydailynews.com/opinion/ny-oped-educate-kids-in-our-schoolyards-20200729-ym77agselbbzflt4byknh6yf5q-story.html

[8] Prakash Nair. The Classroom is Obsolete, It's Time for Something New, Commentary, Education Week, July 29, 2011
https://educationdesign.com/wp-content/uploads/2020/03/The_Classroom_is_Obsolete-Ed-Week.pdf

[9] Green Schoolyards America. Outdoor Spaces Are Essential Assets For School Districts' Covid-19 Response Across The USA. A Proposal to Engage School Grounds and Parks as Strategic, Cost-Effective Tools for Improving Academic, Mental and Physical Wellbeing as Schools Reopen, July 21, 2020 https://www.greenschoolyards.org/

[10] Prakash Nair. The Classroom is Obsolete, It's Time for Something New, Commentary, Education Week, July 29, 2011

[11] This section and the four that follow are based largely on, and extract relevant sections from, the book titled, Blueprint for Tomorrow, Redesigning Schools for Student-Centered Learning by Prakash Nair, Harvard Education Press, 2014
https://www.amazon.com/Blueprint-Tomorrow-Redesigning-Student-Centered-Learning/dp/1612507042

[12] Timothy Egan. Nature Deficit Disorder, New York Times, March 29, 2012. http://opinionator.blogs.nytimes.com/2012/03/29/nature-deficit-disorder/ and
Richard Louv. Last Child in the Woods (Algonquin Books, 2005)

[13] A. F. Taylor, F. E. Kuo, & W. C. Sullivan. Coping with ADD: The surprising connection to green play settings. Environment and Behavior, 33(1) (2001), 54-77

[14] Karen Malone & Paul Tranter. Children's Environmental Learning and the Use, Design and Management of Schoolgrounds, Children, Youth and Environments, 13(2) (2003)

[15] William Crain. How Nature Helps Children Develop. Montessori Life, Summer 2001.

[16] E. Cobb. The Ecology of Imagination in Childhood, New York, Columbia University Press, 1977.

[17] Richard Louv. Childhood's Future, New York, Doubleday, 1991.

[18] Ruth A Wilson. The Wonders of Nature - Honoring Children's Ways of Knowing, Early Childhood News, 6(19). 1997.

[19] Robin Moore. Compact Nature: The Role of Playing and Learning Gardens on Children's Lives, Journal of Therapeutic Horticulture, 8 (1996) 72-82

[20] R Nowak. Blame lifestyle for myopia, not genes. NewScientist, July 10, 2004, 12

[21] Sheridan Bartlett. Access to Outdoor Play and Its Implications for Healthy Attachments. (Unpublished article, Putney, VT, 1996)

[22] Robin C. Moore. The Need for Nature: A Childhood Right, Social Justice https://www.jstor.org/stable/29767032?seq=1 Vol. 24, No. 3 (69), Children and The Environment (Fall 1997), pp. 203-220

[23] Robin Moore. Compact Nature: The Role of Playing and Learning Gardens on Children's Lives, Journal of Therapeutic Horticulture, 8 (1996) 72-82

[24] Miki Perkins (2012) Hey adults, we just want to let our hair down, The Age, Melbourne, Australia, June 9, 2012. http://www.theage.com.au/victoria/hey-adults-we-just-want-to-let-our-hair-down-20120608-201qf.html

[25] Prakash Nair, Roni Zimmer Doctori and Dr. Richard F. Elmore. Learning by Design. Live |Play | Engage | Create, Education Design International, 2020

[26] Urban Seedling, Montreal, CA. https://www.urbanseedling.com/about/

[27] Mallika Nair, Growing Together, http://www.growingtogetherproject.org/

[28] Prakash Nair, Roni Zimmer Doctori and Dr. Richard F. Elmore. Learning by Design. Live |Play | Engage | Create, Education Design International, 2020

[29] Amanda Stutt, Curriculum Development and the Three Models Explained, Top Hat Blog, October 25, 2018 https://tophat.com/blog/curriculum-development-models-design/

[30] I recommend books by Curriculum 21 and Bold Moves by Heidi Hayes Jacobs (primary author) and Invent to Learn by Dr. Gary Stager and Sylvia Lybov Martinez for schools contemplating the outdoors. Many of the ideas presented in these books will be far easier to put into practice in an outdoor setting than within the confines of a traditional classroom.

[31] The National COVID-19 Outdoor Learning Initiative https://www.greenschoolyards.org/covid-19-overview-outdoor-learning

[32] Inside the quest to reopen schools – by moving classes outside by Nate Berg, Fast Company, July 20, 2020 https://www.fastcompany.com/90532401/inside-the-quest-to-reopen-schools-by-moving-classes-outside

[33] For more information about the Outdoor Spaces Assessment APPS mentioned in this paper, please write to info@educationdesign.com or call +1.800.311.2429 or +1.917.406.3120

5 CHAPTER OUTDOOR LEARNING

FIGURE 82. At the American Embassy School in New Delhi, India, an interior courtyard has been transformed into a learning terrace whose micro-climate is moderated by shade and lots of greenery.

FIGURE 83. Consider mini amphitheaters in high-traffic areas.

5 CHAPTER OUTDOOR LEARNING

FIGURE 83/84. While green amphitheaters are the most desirable, more modest ones like these paved examples can also work if they are adequately shaded.

FIGURE 85. Among the many ways in which outdoor areas can be used, a logical one is for casual eating. As with most other outdoor activities, shaded areas work best for outdoor eating.

5 CHAPTER OUTDOOR LEARNING

FIGURE 87. Table surfaces that can be easily cleaned are more suitable for outdoor projects projects. This example also shows how inter-age pairings are easier to do outside of the typical indoor classroom format.

FIGURE 86. There are no real limits on the learning activities that can be conducted outside. Here is an example from Riverside School in India of students doing an outdoor art project.

FIGURE 88. Gardening with children has many benefits. It connects them with nature and helps them breathe fresh air, gets them interested in fruits and vegetables making them more likely to eat them, gets them away from the computer screen, involves physical activity and makes them more environmentally conscious. This picture shows children participating in a tree-planting event sponsored by Growing Together in California.
Photo © Jason Clary. Growing Together: Mallika Nair, Founder.

5 CHAPTER OUTDOOR LEARNING

FIGURE 89/90. *The theoretical work done in the school attains greater meaning when it is based on real-world experiences. Hiking in nature is as authentic as it gets. There are numerous opportunities for students of all ages to acquire a variety of useful skills on a nature walk including teamwork, observation, endurance, and learning about the natural world that is far removed from the screens on their digital gadgets.*

FIGURE 91. *Sometimes the simplest places are also the most appealing, like this space that can be used for a variety of physical fitness activities including dance, yoga and aerobics. Photo: Peter Aaron/Esto*

5 CHAPTER OUTDOOR LEARNING

FIGURE 92. Simple seating arrangements such as the one pictured here are easy to incorporate as part of any school renovation project. Outdoor areas tend to get used a lot more when children have places to sit.

FIGURE 93. Graduation Ceremony at Tampa Prep High School in Tampa, Florida. Social distancing and the requirement that all attendees wear masks made this ceremony held in the school's football field possible. Many such adjustments are being made by schools as they learn to live safely with COVID-19.

FIGURE 94. This outdoor café works because it is under a lightweight awning. Shade structures like this are easy to create and can be quickly installed to increase the amount of outdoor space that is available for use during most of the school day.

5 CHAPTER OUTDOOR LEARNING

FIGURE 95. At NIST International School in Bangkok, Thailand, in the background a shade structure is visible that provides protection from driving winds and rain, which is important in a tropical country like Thailand.

FIGURE 96. This is the play area at NIST below the shade structure which is open on all sides to allow it to be sufficiently daylit and more comfortable than the areas under direct sun.

FIGURE 97. Outdoor spaces that may have been uninhabitable due to heat and humidity are used here to create family-style groupings of students in small gazebos. Fans keep the air moving and allow such spaces to be used all day long.

5 CHAPTER OUTDOOR LEARNING

FIGURE 98. This is a greenhouse built as a geodesic dome in BC, Canada. It is a quick way to cover an area to make it habitable all year and a fun place for children to learn in. The geodesic dome itself is a fascinating engineering subject for study.
Photo: Donna Hausken

FIGURE 99. The beauty of the greenhouse is that it is a space for year-round use. It is a great way to get children to put their technology aside and be active outside working with natural materials. Greenhouses like this one are relatively inexpensive and can be installed quickly, making good use of previously unutilized outdoor areas.

FIGURE 100. Small outdoor sheds like this are healthy to be out in, attractive and easy to build. They are also great places for children to meet in small groups with an adult to work on hands-on assignments. We do not recommend simply using such places as typical "outdoor classrooms."

HAPPY OUTDOOR LEARNING!

Ar. Paras Sareen
Dr Parul Minhas

URBAN SCHOOL GREENING

Stimulating Cognitive Growth and Fostering Environmental Stewardship

Introduction

In an increasingly urbanized and technology-dependent society, the advantages of children interacting with nature are being recognized not just as beneficial but essential. Scientific research affirms the multifaceted gains—physical, mental, psychological, and cognitive—that natural settings can bestow upon young learners. Despite this understanding, urban schools often find themselves grappling with the practical challenge of limited space, which hampers their ability to create natural learning environments. Faced with this dilemma, innovative and flexible solutions become not just desirable, but necessary. This chapter delves into creative approaches such as vertical and rooftop gardens, adaptive use of underutilized spaces like driveways, incorporation of adventure elements, and the creation of miniature 'Miyawaki' forests to overcome these spatial limitations. Our aim is to equip urban educational institutions with actionable strategies that not only enrich the learning environment but also foster cognitive development, boost self-confidence, and nurture a sense of environmental stewardship in students.

Urban Vertical Gardens: Compact Green Wonders

An ingenious solution for space-constrained urban schools is the creation of vertical gardens. These installations, which can inhabit semi-open spaces like verandas or small courtyards, bring a touch of nature into the heart of the school environment. These compact natural walls are an efficient way to screen off building related services. With natural light and ventilation, they can flourish in a wide spectrum of climatic conditions. Particularly, when planted with region-native or adaptive species, vertical gardens require minimal maintenance, making them ideal for schools. These green walls serve as interactive learning spaces, where students' involvement in their upkeep sharpens cognitive abilities, providing insights into cause-and-effect relationships.

FIGURE 101 (a). Lush green walls breathe life into the surroundings, creating an oasis for learning.

FIGURE 101 (b). *A touch of nature elevating ambience, effortlessly blending aesthetics, and sustainability.*

Rooftop Gardens: Sunlit Spaces for Learning

Rooftops provide an untapped potential for urban greening. With abundant sunlight, these flat spaces can become thriving gardens, given proper drainage and waterproofing. Urban schools can take this opportunity a step further by introducing organic farming and hydroponics. Growing fruits and vegetables on rooftops not only aids in students' understanding of sustainable agriculture but also fosters a sense of accomplishment as they witness the results of their diligent care.

FIGURE 102 (a). *Rooftop Gardens as an Enchanting, sustainable learning space amidst the vibrant embrace of nature.*

FIGURE 102 (b). *Seizing the Urban Rooftop Opportunity - Transforming rooftops into flourishing havens of nature and sustainability.*

Green Pathways and Nooks: Reimagining Unused Spaces

Fire tender paths and driveways often lie underutilized in school premises. By introducing buffer plantations or ground covers on grasscrete surfaces, these spaces can transform into green pathways or nooks. The act of caring for these spaces and observing the growth instills cognitive skills and self-confidence among students. Moreover, this contributes to students' understanding of their role in maintaining an environmentally conscious campus.

FIGURE 103. Nature's intervention into driveways and walkways around the school campus.

Adventure Elements: Nature's Outdoor Learning Spaces

Schools blessed with larger trees can introduce adventure elements such as tree houses and rope climbing. These activities bring students closer to natural environments and have positive impacts on their physical and mental health. Simultaneously, these initiatives foster problem-solving abilities and teamwork, enhancing cognitive growth in a fun, engaging manner.

FIGURE 104. *Natural environment cultivating curiosity and growth among young minds.*

'Miyawaki' Forests: Small-scale Biodiversity

The concept of 'Miyawaki' forests, involving high-density planting of native species, offers an innovative way for even smaller schools or daycare facilities to embrace Biophilic Design. The process of creating and nurturing these miniature forests offers an intimate lesson on biodiversity, emphasizing the importance of local flora. It also acts as an effective tool to foster environmental stewardship among students.

FIGURE 105. *A journey through lush greenery fostering inquisitiveness, and a deeper connection with the natural environment.*

Conclusion

In a world increasingly detached from the natural environment, it is imperative to rethink the way we design and operate urban educational spaces. The challenge of limited outdoor space in urban schools is not insurmountable; rather, it provides an opportunity for innovative solutions that integrate nature into the very fabric of the learning environment. As detailed in this chapter, the implementation of vertical gardens, rooftop gardens, green pathways, adventure elements, and even small-scale 'Miyawaki' forests can significantly enrich the educational experience. These initiatives offer more than just aesthetic appeal; they serve as living classrooms that stimulate cognitive growth, enhance self-confidence, and instill a deep sense of environmental stewardship in students.

The benefits of these green interventions extend beyond the individual to enrich the school community as a whole. By adopting these approaches, schools can transform their campuses into thriving ecosystems that are not only sustainable but also deeply nurturing. Educators and administrators are equipped with a range of actionable strategies to counter the limitations imposed by urban landscapes, providing students an oasis of learning that is as engaging as it is beneficial. Thus, in taking these steps, schools not only enrich their immediate environment but also contribute to a more sustainable, conscientious, and ecologically-responsible future.

References:

1. Beatley, T. (2011). Biophilic cities: integrating nature into urban design and planning. Island Press.

2. Kellert, S. R., Heerwagen, J. H., & Mador, M. L. (Eds.). (2008). Biophilic design: the theory, science, and practice of bringing buildings to life. John Wiley & Sons.

3. Akpinar, A. (2017). How is quality of urban green spaces associated with physical activity and health? Urban Forestry & Urban Greening, 20, 76-83.

4. Li, D., & Sullivan, W. C. (2016). Impact of views to school landscapes on recovery from stress and mental fatigue. Landscape and Urban Planning, 148, 149-158.

5. Wells, N. M. (2000). At home with nature: Effects of" greenness" on children's cognitive functioning. Environment and Behavior, 32(6), 775-795.

6. Miyawaki, A. (1999). Restoration of urban green environments based on the theory of vegetation ecology. Ecological Engineering, 14(1-2), 185-200.

7. Kweon, B. S., Sullivan, W. C., & Wiley, A. R. (1998). Green common spaces and the social integration of inner-city older adults. Environment and Behavior, 30(6), 832-858.

Dr. Parul Minhas
Francesco Cupolo

A Framework for School Design

Introduction

In the evolving landscape of education, understanding the intrinsic needs of learners is paramount. At the intersection of psychology and architectural design lies an exciting potential to enhance educational experiences. Self-Determination Theory (SDT), a framework centered on the fundamental psychological needs, offers invaluable insights in this regard. These needs, while universal, hold significant implications for the design and orchestration of learning environments.

Background of Self-Determination Theory (SDT)

Self-Determination Theory (SDT) is a psychological framework that posits three essential needs as the key drivers of human motivation: autonomy, competence, and relatedness. Developed by psychologists Richard Ryan and Edward Deci, SDT emphasizes the importance of fulfilling these innate psychological needs for optimal development, performance, and well-being.

- **Autonomy** refers to the need to have control over one's actions and choices, and a feeling of volition and self-endorsement in one's behavior.
- **Competence** deals with the drive to interact effectively with one's surroundings and to experience mastery and success in one's endeavors.
- **Relatedness** emphasizes the need to feel connected to others, understood, and part of a community or group.

SDT has been applied across various domains, including education, workplace, healthcare, and sports, among others. In the context of education, SDT principles have guided the development of teaching methods, curricular design, and the creation of learning environments that foster student motivation and success.

The Interplay of Psychology and Architectural Design in Education

The design of learning spaces goes beyond mere aesthetics and functionality. It involves a nuanced interplay between psychology and architecture, where understanding the mental and emotional needs of learners becomes central to the spatial design.

The integration of SDT in the design of educational spaces translates psychological principles into tangible architectural considerations. By tailoring the design to meet the three core needs of autonomy, competence, and relatedness, architects and educators can create environments that not only look appealing but also facilitate positive learning experiences.

- **Autonomy** can be supported by designing flexible and adaptable spaces, giving students the freedom to choose how and where they learn.
- **Competence** is nurtured by creating spaces that enable focus, mastery, and exploration, considering factors like ergonomics, lighting, and acoustics.
- **Relatedness** is fostered by building spaces that enhance social interaction and collaboration, creating a sense of belonging and community.

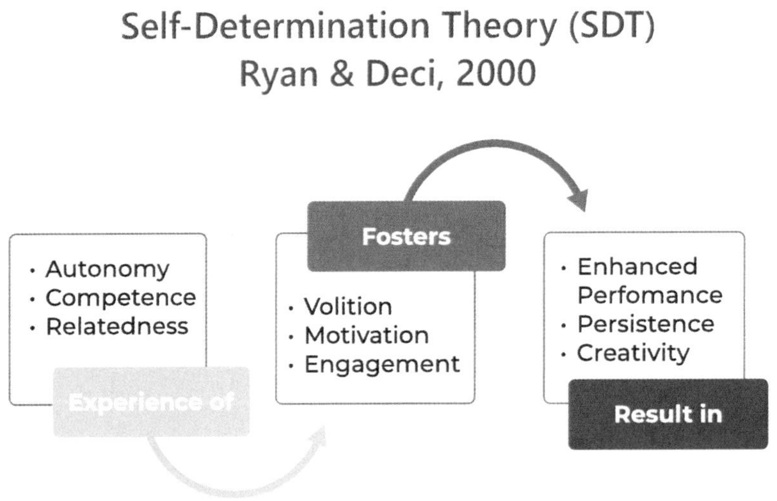

Self-Determination Theory (SDT)
Ryan & Deci, 2000

FIGURE 106. Self-Determination Theory.

This innovative approach transcends traditional design paradigms, allowing learning spaces to resonate with the innate psychological needs of the learners. As the field of education continues to evolve, embracing the principles of SDT in architectural design paves the way for more responsive, empathetic, and effective learning environments.

Understanding Self-Determination Theory (SDT)

The heart of Self-Determination Theory (SDT) lies in its exploration of human motivation, emphasizing the conditions that foster the true nature and potential of individuals. As a macro-theory of human motivation, SDT provides insights into the dynamic interplay between individuals and their environments, shedding light on the optimal conditions required for personal growth, psychological health, and well-being.

FIGURE 107. The role of need satisfaction in motivation according to self-determination theory.

The Principle of Autonomy

Autonomy is the innate need to perceive oneself as the source of one's own behavior. It's the ability to act in harmony with one's self, rather than feeling forced or obligated. This principle doesn't necessarily mean independence or self-sufficiency. Instead, it emphasizes the need to feel volitional and to have the experience of choice.

In educational settings, autonomy is closely associated with:

- **Motivation:** Students who perceive their actions as self-determined are more likely to feel engaged and motivated.
- **Learning outcomes:** Autonomy has been correlated with deeper and more efficient learning, as well as better retention.
- **Well-being:** Feeling autonomous contributes to overall psychological well-being, reducing feelings of stress and burnout.

The Drive of Competence

Competence refers to the need to be effective in one's interactions with the environment and to seek out and master optimal challenges. It's about feeling confident and capable in one's abilities.

In the context of education, the drive for competence manifests as:

- **Engagement:** Students are more likely to engage with materials and activities that they feel competent in.
- **Persistence:** The belief in one's competence can lead to increased persistence in challenging tasks, even in the face of setbacks.
- **Quality of learning:** Feeling competent often leads to a deeper understanding of the material, as students are motivated to understand rather than just memorize.

FIGURE 108. Promoting autonomy in learning requires educators to recognize and nurture students' inherent interests and provide opportunities for them to take the initiative in their learning journeys.

7 CHAPTER SELF-DETERMINATION THEORY

FIGURE 109. Teachers can foster competence in students by providing optimal challenges—tasks that are neither too easy nor too hard—and offering constructive feedback that focuses on effort and strategies rather than innate ability.

The Need for Relatedness

Relatedness speaks to the universal want to interact with, be connected to, and experience caring for others. It's about feeling understood, valued, and having a sense of belonging to a community.

In educational terms, relatedness has implications for:

- **Classroom dynamics:** A sense of relatedness can lead to more cooperative and harmonious classroom interactions.
- **Engagement:** Students are more likely to engage in activities where they feel a connection with their peers or teachers.
- **Mental health:** A feeling of connectedness can be protective against feelings of isolation, loneliness, and depression.

7 CHAPTER SELF-DETERMINATION THEORY

FIGURE 110. To nurture relatedness, educators need to build positive relationships with students and foster peer connections. Architects, in turn, need to create learning spaces where every student feels seen, heard, and valued.

In sum, the three foundational principles of SDT—autonomy, competence, and relatedness—are deeply interconnected, each playing a crucial role in shaping human behavior and motivation. In the realm of education, understanding and addressing these needs can lead to more vibrant, engaged, and successful learning environments.

Spatial Implications Of Self-Determination Theory

Designing for Autonomy

Flexible Learning Spaces and Environmental Orderliness

In order to promote learner autonomy, it's essential to provide flexible learning spaces that allow for a seamless shift between different learning modalities. This entails incorporating movable furniture, modular structures, and intuitive layouts that cater to individual and collective needs. A crucial facet of this is environmental orderliness.

By minimizing clutter and chaos, learning spaces can be designed to be more legible and predictable, reinforcing a sense of control among learners. With intuitive wayfinding elements such as clear landmarks and visual cues, students can navigate their environment effortlessly, further empowering their sense of autonomy.

7 CHAPTER SELF-DETERMINATION THEORY

FIGURE 111. A crucial facet of designing for autonomy is environmental orderliness.

FIGURE 112. Each space can forge its own unique identity through color which also assists with legibility and wayfinding.

FIGURE 113. Since not all students learn in the same way or are comfortable in the same place, it is important to provide a variety of learning spaces. Such variety also allows for learning itself to be variegated and helps student develop social, emotional and collaboration skills that more traditional classroom layouts do not adequately support.

Personal Spaces, Accessibility, and Nature Integration

Each learner is unique, bringing with them diverse needs and preferences. Recognizing this, it's essential to offer personal spaces that respect students' needs for privacy and territoriality. Designing a 'home base' for learners, where they can keep their belongings or return to for quiet reflection, reinforces a sense of personal ownership. Further, to promote inclusivity, a barrier-free design approach is pivotal. From light switches and door handles to water fountains and views through doors/windows, every element should be scaled according to the developmental needs of learners. Integrating nature, be it through direct access or visual pathways, can foster sensory orientation and provide a restorative ambiance, further promoting autonomy.

FIGURE 114. *Giving children access to natural elements, especially in ways that teach them essential skills like gardening is a powerful way to build autonomy.*

7 CHAPTER SELF-DETERMINATION THEORY

FIGURE 115. The more children can do for themselves and the less they need to call upon adults for assistance, the more likely that they will practice autonomy and gain in confidence to navigate their world.

Catering to Competence

Restorative and Interaction Environments

For learners to truly excel and hone their competence, the environment should provide spaces that allow them to restore their attention and focus. This can be achieved by incorporating quiet rooms for breaks and relaxation, as well as green spaces that can serve as sanctuaries amidst the hustle and bustle of learning.

Furthermore, interactions can be crucial for competence-building. Spaces that encourage collaboration, discussions, and peer-to-peer engagement cannot only enhance learning outcomes but also boost confidence. To support this, attention to ambient factors such as lighting, acoustics, indoor air quality, ventilation, and thermal comfort becomes paramount.

FIGURE 116. *Interesting getaway spaces like this treehouse will naturally draw students to them and serve not only as a source of emotional and mental relaxation but also improve their overall health and well-being through their connection with nature.*

7 CHAPTER SELF-DETERMINATION THEORY

FIGURE 117. Access to daylight and views to nature have proven benefits in terms of increasing children's mental agility and also their concentration and efficiency. These benefits are in addition to the obvious health benefits afforded by daylight and visible connections to natural elements.

FIGURE 117. The concept of "comfort" goes beyond physical elements like ergonomic furniture. It also extends to auditory comfort which can be enhanced through the strategic placement of acoustical treatment throughout the learning space – especially more open spaces like the one pictured here.

Prioritizing Relatedness

Inclusive, Collaborative, and Sensory Design

Creating an environment where every learner feels they belong is at the heart of fostering relatedness. This sense of belonging can be enhanced by designing smaller learning communities or introducing intimate spaces within larger institutions. Materials play a significant role in this, with the incorporation of natural, authentic design elements taking precedence over synthetic, decorative ones.

FIGURE 118. A sense of belonging can be enhanced by designing smaller learning communities or introducing intimate spaces within larger institutions.

The colors, textures, patterns, and atmospheric qualities they bring can profoundly affect learners' sense of relatedness. Moreover, dedicated spaces for group activities and incidental social interactions, both among peers and with facilitators, can nurture a deep sense of community.

FIGURE 119. The use of natural materials, particularly those that are indigenous to the local community fosters relatedness and builds community spirit. This space shows a feature wall constructed with pieces of trees that were cut down during the construction of the school.

7 CHAPTER SELF-DETERMINATION THEORY

FIGURE 120. *Students will naturally gravitate to spaces like the one shown above where socializing or collaborating in groups is not just facilitated, but encouraged as well.*

Nature Connectedness in Design

Nature has a unique way of connecting individuals not only to the environment but also to one another. By designing spaces that offer direct interactions with nature and other life forms, learners can develop a profound sense of relatedness. Whether it's through plant-filled nooks, outdoor learning spaces, or large windows that offer panoramic views of nature, integrating natural elements can stimulate a deep sense of connection and belonging.

By integrating these considerations, architects can create learning environments that holistically cater to the intrinsic needs of learners, as highlighted by the Self-Determination Theory.

FIGURE 121. *Natural elements, especially those that feature water, become places where students are more likely to congregate for social and academic activities.*

TABLE 7. Spatial Implications of SDT for Learning Spaces.

SDT Principle	Design Focus	Architectural Highlights & Considerations
Autonomy	Flexible Learning Spaces	Movable furniture for adaptability, Density control/personal space, Variety of learning spaces
	Environmental Orderliness & Predictability	Elimination of chaos, Intuitive wayfinding & landmarks, Legibility and visual order
	Personal Spaces & Accessibility	Home base for learners, Access to private rooms and territoriality, Barrier-free design, Scale-based design (switches, rails, steps, views)
	Natural Integration	Access to nature/daylight sensory orientation, Views through doors/windows
Competence	Restorative Spaces	Spaces for attention restoration/concentration, Quiet rooms for breaks and relaxation
	Environments for Interaction	Provision of interaction spaces/green spaces, Attention to ergonomics, lighting, acoustics, air quality, ventilation, thermal comfort, colors, ambiance
	Accessibility	Barrier-free environment
	Small Learning Communities	Create smaller schools or communities within larger ones
Relatedness	Sensory Design	Engage senses: color, texture, pattern, light, temperature, sound, Authentic design elements and near-nature materials over decorative and synthetic ones
	Collaborative & Social Spaces	Spaces for group activities, Informal, incidental social interactions, Spaces that foster interactions among peers and between students and facilitators
	Nature Connectedness	Engaging natural environments, Opportunities for interaction with nature and other life forms (plants and animals)

Challenges And Considerations For The Future

Addressing Diverse Student Needs

Understanding and catering to the varied needs of learners is one of the primary challenges in designing spaces based on SDT. Every student comes from a unique background, has distinct learning preferences, and is at different stages of their learning journey. Creating spaces that are versatile enough to meet these diverse requirements is paramount. Strategies include understanding cultural nuances, making provisions for differently-abled students, and allowing for adaptable spaces that can be tailored to evolving pedagogies.

Balancing SDT Principles with Practical Constraints

As with any architectural endeavor, practical constraints, whether budgetary, spatial, or regulatory, can impact the design process. It's a challenge to balance the ideals of SDT with the realities of construction, maintenance, and operations. This may involve prioritizing certain SDT principles over others based on the immediate context, considering long-term sustainability over short-term achievements, or innovatively optimizing available resources to meet as many SDT principles as feasible.

7 CHAPTER SELF-DETERMINATION THEORY

The Rise of Virtual Learning Environments

The education landscape has been undergoing a seismic shift with the rise of virtual and hybrid learning models. This transition brings with it a set of challenges in integrating SDT principles. While online platforms offer autonomy, the feeling of competence and relatedness can be harder to cultivate in a virtual space. Designing these environments requires an understanding of digital tools, user experience design, and the dynamics of online engagement to ensure

Conclusion and Future Directions

The intersection of SDT and architectural design in learning spaces is a dynamic field, continuously evolving with pedagogical shifts, technological advancements, and societal changes. While the principles of SDT remain consistent, their application in design will inevitably undergo transformations.

As we move forward, a few considerations become evident. First, the role of technology in shaping learning environments, both physical and virtual, will be paramount. Second, the emphasis on holistic well-being, integrating mental, emotional, and physical health aspects into design, will grow stronger. Lastly, as the global education landscape becomes more interconnected and diverse, designing for inclusivity and equity will be more crucial than ever.

The journey of integrating SDT into architectural design promises to be one of innovation, introspection, and immense potential. By anchoring our designs in the intrinsic needs of learners, we pave the way for more engaged, motivated, and fulfilled students, shaping the educational landscapes of the future.

References:

Bandura, A. (1989). Self-efficacy. In V. S. Ramachandran (Ed.), Encyclopedia of human behavior (Vol. 4, pp. 71-81). New York: Academic Press.

Benware, C. A., & Deci, E. L. (1984). Quality of learning with an active versus passive motivational set. American Educational Research Journal, 21(4), 755-765.

Deci, E. L., & Ryan, R. M. (2000). The "what" and "why" of goal pursuits: Human needs and the self-determination of behavior. Psychological Inquiry, 11(4), 227-268.

Deci, E. L., Vallerand, R. J., Pelletier, L. G., & Ryan, R. M. (1991). Motivation and education: The self-determination perspective. Educational Psychologist, 26(3-4), 325-346.

Legault, L. (2017). Self-Determination Theory. In: Zeigler-Hill V., Shackelford T. (eds) Encyclopedia of Personality and Individual Differences. Springer, Cham

Lowenstein, G. (1994). The psychology of curiosity: A review and reinterpretation. Psychological Bulletin, 116(1), 75-98.

Montessori, M. (No Date). The Montessori method. (No Publisher).

Ryan, R. M. (1995). Psychological needs and the facilitation of integrative processes. Journal of Personality, 63(3), 397-427.

Ryan, R. M., & Deci, E. L. (2000). Self-determination theory and the facilitation of intrinsic motivation, social development, and well-being. American Psychologist, 55(1), 68-78.

Ryan, R. M., & Grolnick, W. S. (1986). Origins and pawns in the classroom: Self-report and projective assessments of individual differences in children's perceptions. Journal of Personality and Social Psychology, 50(3), 550-558.

Silvia, P. J. (2008). Interest—The curious emotion. Current Directions in Psychological Science, 17(1), 57-60

Prakash Nair, AIA
Dr. Parul Minhas

UNMASKING PSEUDOTEACHING

Empowering Students Through Active and Authentic Learning

Introduction

Education is a powerful tool that shapes the future of societies. Teachers, as well as architects, play crucial roles in this process. Their primary goal is to facilitate genuine learning, foster critical thinking, and nurture intellectual curiosity. However, in some instances, educators unknowingly engage in a practice known as pseudoteaching with architects as unwitting accomplices.

Pseudoteaching, a term coined by Frank Noschese, refers to well-intentioned teaching methods that appear effective on the surface but fail to promote deep understanding and long-term retention of knowledge. Noschese's work highlights the importance of critically examining teaching practices to ensure they truly promote deep comprehension and meaningful learning. This article delves into the concept of pseudoteaching, explores its relationship with the traditional classroom-based model of schooling, and advocates for a shift in the design of learning spaces to facilitate an authentic and student-directed approach to learning.

The Deceptive Facade Of Pseudoteaching

FIGURE 122. Teaching methods like this form of direct instruction may appear effective on the surface but fail to promote deep understanding.

FIGURE 123. A big lecture like this one, even from an expert, may only result in superficial familiarity as opposed to deep understanding of what is taught.

Pseudoteaching is akin to a magician's trick, where the illusion of learning captivates students and teachers alike. Educators often employ techniques that produce impressive short-term results, such as memorization drills, rote learning, and excessive teacher-led instruction. Students may regurgitate information accurately in the short term, leading both teachers and students to believe that learning has occurred. However, this superficial understanding often crumbles when faced with real-world application or the need for critical thinking.

The Roots Of Pseudoteaching

Several factors contribute to the prevalence of pseudoteaching in classrooms today. One significant factor is the emphasis on standardized testing and the pressure to meet predetermined academic benchmarks. This focus can lead teachers to prioritize teaching to the test rather than fostering deep comprehension. Additionally, the use of traditional teaching methods, such as lecturing, worksheets, and memorization, may be deeply ingrained in educational systems. These methods are often convenient and familiar, **but they limit students' active engagement and critical thinking**. Furthermore, the classroom-based model of schooling reinforces pseudoteaching by its very design.

8 CHAPTER UNMASKING PSEUDOTEACHING

FIGURE 124. *The classroom-based model of education dates to the late 19th century and was designed for the delivery of content and not the development of modern-day skills like complex problem solving and creativity.*

FIGURE 125. *One significant factor contributing to the prevalence of pseudoteaching is the emphasis on standardized testing.*

Unveiling The Pseudoteaching Paradox

Pseudoteaching thrives within the confines of traditional classroom-based education, where the focus often lies on passive rather than active learning. Classrooms are typically designed to promote teacher-centered instruction, with students assuming a passive role as recipients of knowledge. **This passive learning approach**, characterized by rote memorization, regurgitation, and compliance, **reinforces the illusion of learning** without fostering true understanding. As a result, students may struggle to apply knowledge in real-world situations or engage in critical thinking.

FIGURE 126. Classrooms are typically designed to promote teacher-centered instruction.

FIGURE 127. Rearranging the classroom into table groupings can reduce pseudoteaching but the spaces are still inadequate for differentiation and hands on learning.

8 CHAPTER UNMASKING PSEUDOTEACHING

The Need For A New Direction

To address pseudoteaching effectively, there is a need for a new direction in education that also incorporates insights from the field of architecture. The traditional classroom-based model must evolve to prioritize student-centered learning approaches, while also considering the physical design of learning environments. This paradigm shift involves moving away from passive learning and embracing active engagement, authentic experiences, and student-directed learning. By integrating principles of spatial design and considering factors such as variety and flexibility of spaces, we can create learning environments that support collaborative and interactive learning.

FIGURE 128. *This is a learning suite where two learning studios are combined to increase opportunities for student-centered learning.*

FIGURE 129. *This space discourages pseudoteaching by allowing for multiple modalities of learning to occur simultaneously.*

8 CHAPTER UNMASKING PSEUDOTEACHING

Fostering Authentic And Hands-On Learning

In the pursuit of authentic learning experiences, the physical design of the environments in which children learn must be reimagined to accommodate active and hands-on learning. Architects can work closely with educators to create dynamic and flexible environments that encourage collaboration, creativity, and exploration. By integrating project-based learning, problem-solving activities, and real-world applications, students can gain a deeper understanding of concepts and develop vital skills such as critical thinking, communication, and adaptability. **Moreover, hands-on experiences**, such as experiments, field trips, and community engagement, **provide opportunities for students to connect theory with practice**, enhancing their comprehension and fostering a love for lifelong learning.

FIGURE 130. *Learning areas should be designed for a variety of student-led projects (preferably with outdoor connections) to help children develop collaboration and hands-on problem-solving skills.*

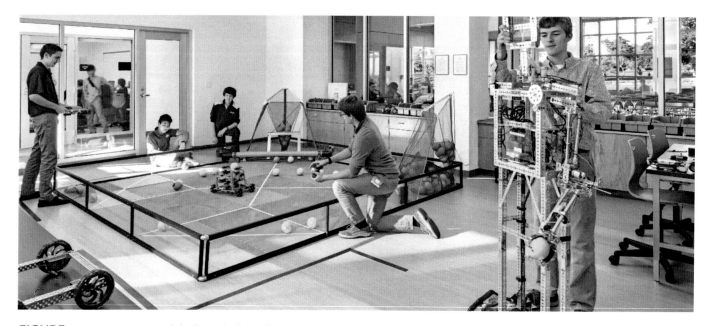

FIGURE 131. *Dispensing with hallways allows for the design of open spaces that can be quickly deployed for experiments that need large open spaces — experiments that would be impossible in traditional classrooms.*

FIGURE 132. Outdoor activities such as this student-created vegetable garden provide authentic learning experiences that are more meaningful than the most engaging classroom lecture.

FIGURE 133. Even traditional "indoor" games like chess come alive when it becomes an outdoor activity.

8 CHAPTER UNMASKING PSEUDOTEACHING

The Role Of Teachers And Architects

In an active learning environment, teachers serve as facilitators and guides, while architects play a crucial role in designing spaces that support effective teaching and learning. Teachers encourage inquiry, pose thought-provoking questions, and guide students to discover answers through their own exploration. They create a safe and supportive space for students to take risks, make mistakes, and learn from their experiences. Architects, on the other hand, consider the physical elements of the learning environment, ensuring that spaces are adaptable, aesthetically pleasing, and conducive to collaboration and engagement.

FIGURE 134. Learning Spaces that free teachers from the bane of pseudoteaching allows them to assume the role of mentors.

FIGURE 135. Learning commons are flexible spaces where teachers can provide as-needed help to students that need it.

Breaking Free From Pseudoteaching

To overcome pseudoteaching and embrace student-centered learning, educators, architects, policymakers, and stakeholders must collaborate to reimagine and redesign the educational landscape. This transformation involves providing professional development opportunities for teachers to acquire the necessary skills and knowledge to implement student-directed learning effectively. Additionally, investments in educational resources, infrastructure, and technology are essential to create supportive environments that facilitate authentic and hands-on learning experiences. By prioritizing students' agency, curiosity, and individual growth, we can empower them to become lifelong learners and active contributors to society.

FIGURE 136. *A computer lab is a perfect metaphor for the prevalence of pseudoteaching. The computer connects each student to the entire universe of information and knowledge and yet they are placed in uniform rows and children are "taught" by one individual who tells them exactly what to do.*

FIGURE 137. *The same lab pictured above was converted to this "innovation lab" where technology is ubiquitous, but children are in charge.*

8 CHAPTER UNMASKING PSEUDOTEACHING

FIGURE 138. *This is how the power of pedagogy, curriculum and architecture can combine to replace pseudoteaching with authentic, teacher-guided, student-directed learning.*

Conclusion

Pseudoteaching is a persistent challenge within traditional classroom-based education, but it is not insurmountable. By recognizing the limitations of passive learning, incorporating insights from architecture, and embracing authentic, student-directed approaches, we can dismantle the illusion of learning and foster deep comprehension. It is time for a paradigm shift—one that redefines the role of teachers as guides, learning environments as dynamic spaces, and students as active participants in their educational journeys. Let us seize this opportunity to transform education, equipping students with the skills, knowledge, and mindset they need to thrive in an ever-evolving world. The time for action is now.

Comparison of Old and New Paradigm of Teaching
(Johnson, Johnson & Smith, 1991)

	Old Paradigm	**New Paradigm**
Knowledge	Transferred from Faculty to Students	Jointly Constructed by Students and Faculty
Students	Passive Vessel to be filled by Faculty's Knowledge	Active Constructor, Discoverer, Transformer of Knowledge
Faculty Purpose	Classify and Sort Students	Develop Students' Competencies and Talents
Relationships	Impersonal Relationships Among Students and Between Faculty and Students	Personal Transaction Among Students and Between Faculty and Students
Context	Competitive/Individualistic	Cooperative Learning in Classroom and Cooperative Teams Among Faculty
Teaching Assumption	Any Expert can Teach	Teaching is Complex and Requires Considerable Training

Table 1. Johnson, D.W., Johnson, R.T., and Smith, K.A. Active Learning: Cooperation in the College Classroom (1st ed.). Edina, MN: Interaction Book Company, 1991.

References:

1. Bligh, D. A. (2000). What's the Use of Lectures? Jossey-Bass.

2. Deslauriers, L., Schelew, E., & Wieman, C. (2011). Improved learning in a large-enrollment physics class. Science, 332(6031), 862-864.

3. Dunlosky, J., Rawson, K. A., Marsh, E. J., Nathan, M. J., & Willingham, D. T. (2013). Improving students' learning with effective learning techniques: Promising directions from cognitive and educational psychology. Psychological Science in the Public Interest, 14(1), 4-58.

4. Hake, R. R. (1998). Interactive-engagement versus traditional methods: A six-thousand-student survey of mechanics test data for introductory physics courses. American Journal of Physics, 66(1), 64-74.

5. Hattie, J. (2009). Visible Learning: A Synthesis of Over 800 Meta-Analyses Relating to Achievement. Routledge.

6. Johnson, D. W., Johnson, R. T., & Smith, K. A. (1998). Active Learning: Cooperation in the College Classroom. Interaction Book Company.

7. Kirschner, P. A., Sweller, J., & Clark, R. E. (2006). Why Minimal Guidance During Instruction Does Not Work: An Analysis of the Failure of Constructivist, Discovery, Problem-Based, Experiential, and Inquiry-Based Teaching. Educational Psychologist, 41(2), 75-86.

8. Kuhlthau, C. C., Maniotes, L. K., & Caspari, A. K. (2015). Guided Inquiry: Learning in the 21st Century (2nd ed.). Libraries Unlimited.

9. Lemov, D. (2010). Teach Like a Champion: 49 Techniques that Put Students on the Path to College. Jossey-Bass.

10. Mayer, R. E., & Johnson, C. I. (2008). Revising the redundancy principle in multimedia learning. Journal of Educational Psychology, 100(2), 380-386.

11. National Research Council. (2012). Education for Life and Work: Developing Transferable Knowledge and Skills in the 21st Century. National Academies Press.

12. Noschese, F. (2009). Pseudoteaching. Action-Reaction. Retrieved from http://fnoschese.wordpress.com/2009/11/15/pseudoteaching/

13. Prince, M. (2004). Does Active Learning Work? A Review of the Research. Journal of Engineering Education, 93(3), 223-231.

14. Tishman, S., Jay, E., & Perkins, D. (1993). Teaching as the Learning Profession: Handbook of Policy and Practice. Jossey-Bass.

15. Vygotsky, L. S. (1978). Mind in Society: The Development of Higher Psychological Processes. Harvard University Press.

16. Wieman, C. E. (2014). Large-scale comparison of science teaching methods sends clear message. Proceedings of the National Academy of Sciences, 111(23), 8319-8320.

CHAPTER 9

Dr. Parul Minhas
Prakash Nair, AIA

CHOICE ARCHITECTURE

Nudging Behavioral Shifts
for Holistic Well-being

Executive Summary:

In today's rapidly evolving world, schools have transformed from mere infrastructures to dynamic environments that significantly influence students' behavior, health, and futures. Historically, the design of school environments prioritized utility and functionality, often overlooking the subtle but profound impacts of these spaces on students' holistic well-being — an amalgamation of their physical, mental, emotional, and environmental health. Enter Choice Architecture: a groundbreaking approach rooted in behavioral economics that utilizes environmental cues to guide individuals towards beneficial decisions. When integrated into school designs, Choice Architecture promises transformative outcomes. This chapter champions a paradigm shift in school design, urging architects, educators, policymakers, and stakeholders to adopt Choice Architecture. By fusing architectural expertise with behavioral insights, we can reconstruct learning environments that nurture not just academic growth, but also holistic well-being.

Choice Architecture - A New Lens For School Design

FIGURE 139. *Crafting choices for the holistic development of children.*

9 CHAPTER CHOICE ARCHITECTURE

In our ever-evolving world, environments play a pivotal role in shaping behaviors, attitudes, and outcomes. From the urban sprawl of bustling cities to the digital spaces we frequent online, every setting subtly nudges us toward specific choices. Schools, which act as a crucible for early development and deeply impact our future generations, are no exception. Their designs, often taken for granted, wield an enormous influence on a student's holistic health - spanning physical, mental, emotional, and environmental dimensions.

The traditional approach to designing schools has largely been predicated on static, utilitarian principles: assembly halls here, libraries there, playgrounds outside. A sort of formulaic blueprint, efficient but often lacking in terms of fostering holistic health and adaptability. However, what if there were a way to bridge the conventional methodologies of school design with insights from behavioral science? **What if our school environments could not only accommodate but also subtly nudge students towards better health, community spirit, and environmental stewardship?**

Enter Choice Architecture, a concept hailing from the corridors of behavioral economics, pioneered by thinkers like Richard Thaler and Cass Sunstein. In essence, Choice Architecture revolves around the design of different ways in which choices can be presented to consumers, and its impact on decision-making. It's the art and science of curating environments to inspire better decisions, without forcibly directing them. It's about understanding the nuances of human behavior and molding surroundings to reflect and respond to those intricacies.

The potential of integrating Choice Architecture into school design is vast and largely uncharted. By leveraging its principles, architects and educators can craft spaces that resonate with a student's innate tendencies, preferences, and needs, thus promoting holistic well-being. This chapter delves deep into the world of Choice Architecture, exploring its foundational tenets, and weaving them into the fabric of school design. It presents an argument for a new era of school environments — one where every corner, every open space, and every structure champions the cause of holistic health.

Understanding Choice Architecture

The Foundation: Core Philosophy of Choice Architecture in Educational Settings

The bedrock of Choice Architecture in educational contexts lies in the understanding that decisions are influenced by how choices are presented. It's not just about presenting options; it's about structuring and framing them in ways that facilitate informed decision-making. This approach is centered on creating an environment where students are empowered to make choices that promote their best interests and holistic health.

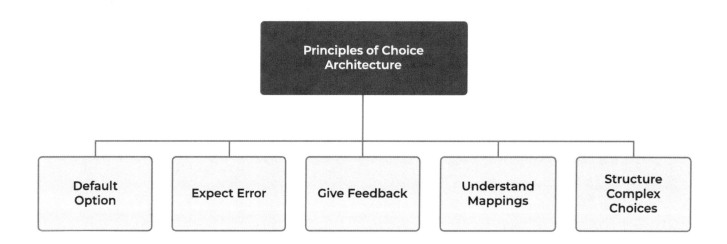

FIGURE 140. Principles of Choice Architecture Tailored for Holistic Health in Schools.

1. **Default Options:** Design school spaces where health-centric and educational choices stand out. For example, ergonomic seating arrangements for better posture or easy access to hydration stations to encourage regular water intake.
2. **Expect Error:** Children, being naturally curious, might err as they explore. Design spaces like labs or playgrounds to be forgiving of these errors. Integrate safety measures and provide clear guidelines, ensuring an environment that balances exploration with safety.
3. **Give Feedback:** Introduce mechanisms that offer children insights into their actions' outcomes. For instance, a garden maintained by students can showcase the results of their care, linking efforts to tangible outcomes.
4. **Understand Mappings:** Ensure that the benefits of certain actions are immediately visible or intuitive in the school's design. For instance, a designated calm corner filled with soft cushions, dim lights, and peaceful imagery can emphasize the connection between quiet moments and mental well-being, prompting students to seek out such spaces when they need to relax or decompress.
5. **Structure Complex Choices:** To support the holistic development of children, break down multifaceted choices into simpler, more intuitive decisions. For instance, a school might introduce a "Personal Growth Hour" where students choose from a variety of activities. One station allows students to journal or draw, promoting self-expression and emotional understanding. Another station might have group puzzles or games, emphasizing teamwork and cognitive development. Yet another could be dedicated to the care of classroom plants or pets, nurturing responsibility and empathy. By segmenting this hour into distinct choices, the school helps students navigate and understand the diverse facets of their development.

Why Is Choice So Important?

Autonomy and Agency in Learning

The bedrock of Choice Architecture in educational contexts lies in the understanding that decisions are influenced by how choices are presented. It's not just about presenting options; it's about structuring and framing them in ways that facilitate informed decision-making. This approach is centered on creating an environment where students are empowered to make choices that promote their best interests and holistic health.

At its core, autonomy is about having control over one's actions and decisions, while agency refers to one's capacity to act in any given environment. For children, having a sense of agency and autonomy in their learning is crucial. Studies have shown that when students feel they have a choice, even if it's between pre-selected options, their intrinsic motivation, engagement, and comprehension increase. This not only fosters academic development but also contributes to their self-worth, confidence, and understanding of self.

The Subtle Power of Nudging

Choice architecture employs "nudges" – subtle design interventions that guide individuals towards specific behaviors without stripping them of their freedom to choose. In a school setting, this could be as simple as placing healthier food options at eye level in a cafeteria or placing student seating closer to areas with more daylight. By making the healthier or more beneficial option more accessible or enticing, you're nudging students towards that choice. However, since they still get to make the final decision, they retain their sense of autonomy.

Holistic Development through Autonomy:

A child's sense of autonomy and agency directly affects his or her holistic development in the following ways:

- **Cognitive:** Choosing their learning path or method can enhance critical thinking and decision-making skills.
- **Emotional:** Making decisions and experiencing their outcomes can foster emotional intelligence, resilience, and coping skills.
- **Social:** Autonomy in interactions aids in developing empathy, understanding boundaries, and improving communication skills.
- **Physical:** Choices related to physical activity or even dietary habits in a school environment can influence long-term health behaviors.

Choice Architecture acknowledges the innate human need for control while ensuring that the choices made align with holistic well-being. By incorporating these principles, educational institutions can cultivate environments where children are nudged towards health, well-being, and success, all while feeling empowered in their journey.

9 CHAPTER CHOICE ARCHITECTURE

FIGURE 141. Choice architecture, when applied with a keen understanding of autonomy and agency is more likely to elicit positive behaviors leading to lasting good habits in students than simple mandates by the adults in school.

Principles Of Choice Architecture In Action For Holistic School Environments

In the evolving landscape of educational institutions, the emphasis has shifted from merely academic performance to a broader perspective of holistic health. This entails nurturing the mental, physical, social, and emotional facets of a student's life. Incorporating Choice Architecture principles into school design can significantly aid in fostering this holistic development. Let's explore how Choice Architecture principles can shape the foundation of holistically designed school environments.

1. Physical Health Facilitation

a) Active Zones: By designing schools with expansive outdoor areas, including playgrounds, sports courts, and green spaces, students are nudged towards physical activity. Similarly, indoor spaces like gymnasiums, dance rooms, or yoga studios offer choices that cater to different physical interests, further promoting active behaviors.

FIGURE 142. Active zones balance exploration with safety.

b) Nutritional Awareness: Canteens and cafeterias can be designed in a way where nutritious food options are more prominently displayed, subtly nudging students towards healthier food choices.

FIGURE 143. *Kitchen gardens where students can grow and harvest their own vegetables promote an understanding and appreciation for fresh produce and makes it more likely that children will be eager to incorporate it into their own diets.*

2. Mental And Emotional Well-Being

a) Quiet Zones: Recognizing the need for students to balance active and restful experiences, it's essential to have strategically placed and attractively designed quiet zones. By making these areas visible and inviting, students are subtly nudged towards areas like reading nooks, meditation rooms, or sensory-deprivation spaces.

FIGURE 144. *Quiet zones offer students opportunities to retreat, find peace, introspect, or take a momentary break.*

9 CHAPTER CHOICE ARCHITECTURE

b) Therapeutic Spaces: Schools must prioritize dedicated spaces for counseling and therapy. Designed with warmth, privacy, and comfort, these rooms not only provide a safe haven for students to address their mental and emotional concerns but also subtly nudge them towards recognizing that seeking help is seen as a sign of strength, not fear.

FIGURE 145. Schools should set aside private, comfortable spaces for counseling that students are more likely to voluntarily use.

3. Social Growth And Interaction

a) Collaborative Spaces: Designed with adaptability in mind, collaborative zones promote teamwork and group activities. The open-concept layout and flexible furniture arrangements allow these areas to transform as needed, encouraging group discussions and project-based learning.

FIGURE 146. Collaborative zones naturally encourage group activities.

b) Cultural Hubs: Thoughtfully designed spaces that spotlight art, music, and cultural diversity are pivotal in promoting holistic growth. By making these areas easily accessible and highly visible, and by adorning them with artworks, musical instruments, and stages, we subtly steer students towards immersive experiences. These environments catalyze creative expression, deepen social connections, and foster an appreciation for the rich tapestry of global cultures.

FIGURE 147. *Its immense versatility makes the black box theatre an obvious hub of student choice for a wide variety of cultural activities.*

4. Environment And Sustainability

a) Eco-awareness Zones: By thoughtfully weaving in elements like botanical gardens or eco-pods into the fabric of school designs, we're not just beautifying spaces but are also subtly guiding students towards environmental consciousness. Positioned prominently, these green zones become places of discovery where students encounter diverse flora and fauna, cultivating a profound sense of respect and responsibility for the environment and grounding them in sustainable practices.

b) Resource Management: Strategically positioning demonstrative features such as rainwater harvesting systems or solar panels not only underlines the school's commitment to sustainability but also acts as a nudge, directing students towards hands-on understanding and appreciation of resource management.

FIGURE 148. *Tangible inclusion of environmentally sustainable solutions drive home the importance of valuing and preserving our natural resources.*

5. Personal Development and Creativity

a) Innovation Labs: Be it in a robotics lab or a DIY craft room, the layout and accessibility of these hands-on areas serve as silent cues, urging students to delve into experimentation, exploration, and discovery.

FIGURE 149. *By intentionally crafting spaces laden with cutting-edge tools and technology, we create an environment that nudges students towards innovation.*

b) Reflection Areas: Schools that encompass more spaces dedicated to self-contemplation, such as areas overlooking nature or quiet terraces, harness the essence of Choice Architecture. Additionally, these serene environments contribute to attention restoration, allowing students to rejuvenate and refocus, further enhancing their personal development.

Integrating these core components into school design, with the underpinning of Choice Architecture principles, can truly pave the way for holistic development. As students navigate through these intentionally designed spaces, they are subtly guided towards healthier choices, enriching experiences, and personal growth.

FIGURE 150. *Areas for reflection that connect with nature offer opportunities for students to engage in introspection, journaling, or simple self-reflection.*

Behavioral Implications Of Applying Choice Architecture In Learning Environments

1. **Enhanced Decision Making:**
   - Rationale: Behavioral economics posits that humans often don't act rationally. Through subtle design cues in learning spaces, we can nudge students towards better decisions.
   - Application: Positioning materials related to pertinent global issues in accessible areas will encourage learners to engage with these topics, broadening their understanding of the world.
2. **Decreased Cognitive Load:**
   - Rationale: An overwhelming number of choices can be paralyzing. Streamlined choices, a principle of choice architecture, can reduce this cognitive burden.
   - Application: Delineating clear zones for varied activities, such as reading and hands-on projects, can simplify decision-making for students, allowing them to engage positively with their environment without feeling inundated.
3. **Motivation and Engagement:**
   - Rationale: Immediate rewards are highly valued in behavioral economics. Spaces offering immediate feedback can be powerful motivators.
   - Application: Tech-integrated areas, where learners instantly visualize the impact of their digital projects, can bolster engagement and motivation.
4. **Encouraging Long-Term Thinking:**
   - Rationale: Delayed gratification is a concept humans grapple with. Thoughtfully designed environments can encourage forward-thinking.
   - Application: Spaces like gardens, where learners can observe the long-term effects of consistent care and effort, can promote patience and the merits of long-term dedication.
5. **Promoting Healthy Risk-Taking:**
   - Rationale: In a well-orchestrated environment, learners are nudged towards beneficial risks, which are indispensable for holistic growth.
   - Application: Areas stocked with diverse creative materials might prompt students to venture into unfamiliar territories, nurturing innovation and adaptability.
6. **Nurturing Empathy and Social Interactions:**
   - Rationale: Organic social interactions can be instrumental in fostering empathy.
   - Application: Common spaces, where learners from diverse backgrounds interact, can encourage a deeper appreciation of varied perspectives.
7. **Instilling a Growth Mindset:**
   - Rationale: Framing, a concept in behavioral economics, shows that the presentation of choices profoundly affects perception.
   - Application: Spaces emphasizing experimentation and iteration can promote a growth mindset, where learners appreciate challenges and view setbacks as learning opportunities.

The principles of choice architecture, when applied to learning environments, hold the potential to significantly influence behaviors. By internalizing these behavioral implications, stakeholders can craft environments that not only facilitate learning but also holistically nurture essential behavioral attributes.

Conclusion

In our comprehensive exploration of Choice Architecture within educational settings, the overarching theme has been the recognition that the way we design and shape our learning environments has profound implications for every facet of a student's well-being — be it mental, emotional, physical, or social.

A unique aspect of our approach, setting it apart from more conventional applications, is its inclusivity. While the primary aim of Choice Architecture is to influence decisions and behaviors for the benefit of the majority, we've widened the lens to ensure every child feels seen and supported. As part of this, we've consciously considered the needs of neurodivergent students. By weaving in features that cater to diverse sensory and cognitive patterns — like quiet zones, adjustable lighting, and flexible seating — we've underscored our commitment to holistic inclusivity.

Furthermore, our philosophy goes beyond the mere physicality of spaces. We champion the idea that learning environments must be adaptable, evolving in stride with both pedagogical shifts and technological advancements. This adaptable ethos not only ensures that the spaces remain relevant but also serves as a manifestation of our dedication to sustainability and inclusivity. The commitment to continuous research and interdisciplinary collaboration is vital. By melding insights from architects, educators, and psychologists, we stay ahead of the curve, ensuring our designs are both innovative and grounded in evidence.

In conclusion, our approach to Choice Architecture in schools isn't just about creating functional spaces; it's about crafting nurturing environments where every student, irrespective of their unique needs or neurotype, can thrive academically and personally. As we look to the future of education, it's clear that the convergence of design and well-being is not just a trend, but a lasting paradigm shift that promises to enrich the learning journeys of countless students.

References & Suggested Readings

Armstrong, T. (2011). *Neurodiversity in the classroom: Strength-based strategies to help students with special needs succeed in school and life.* ASCD.

Barrett, P., Zhang, Y., Moffat, J., & Kobbacy, K. (2013). A holistic, multi-level analysis identifying the impact of classroom design on pupils' learning. *Building and Environment*, 59, 678-689.

Davies, D., Jindal-Snape, D., Collier, C., Digby, R., Hay, P., & Howe, A. (2013). Creative learning environments in education—A systematic literature review. *Thinking Skills and Creativity*, 8, 80-91.

Dweck, C. S. (2006). *Mindset: The new psychology of success.* Random House Incorporated.

Godwin, K., & Seltman, H. (2016). Collaborative learning in small groups: The relationship between workspace design and children's activity patterns. *Journal of Learning Environments Research*, 9(4), 321-339.

Goleman, D., & Senge, P. M. (2016). *The triple focus: A new approach to education.* More Than Sound.

Grandin, T. (2006). *Thinking in pictures: And other reports from my life with autism.* Vintage.

Kellert, S. R. (2005). *Building for life: Designing and understanding the human-nature connection.* Island Press.

Montessori, M. (1967). *The absorbent mind.* Holt, Rinehart and Winston.

Parikh, A., & Parikh, P. (2021). *Choice architecture: A new approach to behavior, design, and wellness.* Routledge.

Parikh, A., & Pati, D. (2022). Choice Architecture and Salutogenesis. In *Choice Architecture and Salutogenesis* (pp. 224-230). Cambridge Scholar Publishing.

Robison, J. E. (2011). *Be different: Adventures of a free-range Aspergian with practical advice for Aspergians, misfits, families & teachers.* Random House.

Thaler, R. H., & Sunstein, C. R. (2008). *Nudge: Improving decisions about health, wealth, and happiness.* Yale University Press.

Tomporowski, P. D., Lambourne, K., & Okumura, M. S. (2011). Physical activity interventions and children's mental function: An introduction and overview. *Preventive Medicine*, 52, S3-S9.

Prakash Nair, AIA

CHAPTER 10: SCHOOL BUILDINGS

The Last Domino

It's Time for Something New

In the post-COVID debate about reopening schools, everyone seemed to agree about one thing: bringing children back to their physical school setting was important. A study published by MIT's Teaching Systems Lab titled, *Imagining September*[i] noted, "One of the most important insights from school closures is the incredible importance of physical school buildings to the work of schools." The study goes on to quote a district leader who said, "building time will be 'gold.'"

I respectfully disagree. Building time has not been "gold" for a long time and, after COVID-19, the fallacy that school buildings serve our children well became that much more evident.[ii] When schools closed abruptly at the start of the pandemic, educators quickly scrambled to offer the same severely limited experience online that they had been offering students in the physical school setting. Most of the evidence that filtered back from "online schools" showed that this model did not work because students were not willing to be held hostage at home in the way they have been in the classroom.

The chart below provides a stark illustration of the extent to which our current model of education is shaped by the school building. The left column describes a defunct educational model that is dictated by the familiar classroom-and-corridor based (cells-and-bells) school building. Taken together, the 13 elements that represent the average school day for millions of children and their teachers describe a model of education far removed from the reality of the world beyond school. Yes, the school building is "incredibly important" but for all the wrong reasons – it prevents schools from delivering the education our children deserve. If so, how can we redesign the school building to facilitate rather than militate against the delivery of a modern education?

In the right column, the change in the physical design of the school from a cells-and-bells to a "learning community" layout can dramatically upgrade all aspects of teaching and learning to make them more relevant for today and tomorrow.[iii]

The good news for schools is that the conversion of a traditional cells-and-bells school into a learning-community based design is relatively inexpensive and can be accomplished quickly. Much can be done within capital budgets already set aside by schools and school districts. Even in cases where a bond referendum is needed to obtain construction money, such efforts become far easier when the facility changes are educationally driven.

One by one, the COVID pandemic upended elements of schooling previously deemed "essential" such as textbooks, standardized tests, mandatory attendance, required number of days in school and homework. In the shadows, the sanitized school building stood waiting to re-exert its socially distanced authority and, in one fell swoop, wipe away all the disruptive innovations its removal from the educational equation had made possible. We do not need to bow to its authority. Today, the traditional school building is the last standing domino preventing the creation of a relevant and effective education system. Knocking it down, metaphorically if not literally, is the critical key to unlocking the incredible creative potential of our children, so that they can go on to build a world far better than the one they are inheriting from us.

TABLE 1. Comparison of the Current Schooling Model to the One that Schools Need to Migrate to.

№	Schooling Dictated by the Cells-and-Bells Model of Physical School Design	Schooling that is Possible in a Learning Community Based Model of School Design
1	Children are forced to spend most of their school day in a classroom	Children choose to occupy spaces best suited to the learning activity they are engaged in
2	Each group of 20 to 30 students is sorted by age and actively supervised by one adult	Students work in multi-age groups of varying sizes, passively supervised by a group of adults
3	In the higher grades, different teachers teach different subjects	A group of teachers working together offers a robust interdisciplinary curriculum
4	The school day is broken down into fixed periods with each period devoted to the study of a different subject	Individual student schedules are organized around the completion of interdisciplinary projects
5	Teachers do not routinely collaborate with peers other than at fixed times set aside for team preparation	Teachers work together in a professional office and collaborate throughout the school day and beyond
6	Direct instruction predominates teaching practice, and even when students are engaged in individual study or teamwork, it is to carry out tasks explicitly defined by the teacher	There is very little direct instruction. Teachers serve as advisors and guide students with the help they need to complete individual or team assignments that students have had a hand in selecting
7	Most of the student assessments measure retention of content and not the demonstration of skills	Individualized formative assessments are designed to provide ongoing feedback and are focused on skill building
8	There is almost no opportunity for teachers to mentor students individually or tutor them in small groups	Teachers can work one-on-one with students, tutor them in small groups, or lecture to larger groups
9	Peer tutoring between students is rare to non-existent	Peer tutoring is encouraged and facilitated by the design of the learning spaces
10	Most of the learning is theoretical, and there is very little hands-on or practical learning	Most of the learning is hands-on and practical with the theory embedded in the assignments
11	Stationary students use mobile technology	Learning spaces amply demonstrate the power of anytime, anywhere learning
12	Students are rarely tapped to express their creativity or engage in complex problem solving – two essential skills for today and tomorrow	Most of the student work is personalized and focused on creativity and complex problem solving

References:

[i] Reich, J. & Mehta J. (2020) Imagining September: Principles and Design Elements for Ambitious Schools during Covid-19, MIT Teaching System Lab. July 3, 2020.

[ii] With all its dysfunction, the physical school is still critically needed for children in underserved communities. For many of these children, school provides a safe haven from abuse and depression, a place to get a healthy meal, and their only access to high-speed Internet. However, a well-designed school is just as important for these children as it is for every student.

[iii] The Learning Community model of school design is fully described in the following two books: 1) Blueprint for Tomorrow – Redesigning Schools for Student-Centered Learning by Prakash Nair, Harvard Education Press, 2014 and 2) Learning by Design. Live | Play | Engage | Create by Prakash Nair, Roni Zimmer Doctori and Dr. Richard F. Elmore, Education Design International, 2020.

Dr. Jeff Melendez
Louis Sirota, AIA

DESIGN FOR SAFETY
Balancing Physical and Psychological Considerations

Introduction

The design of school environments has traditionally leaned heavily on physical safety measures, often overlooking psychological factors that contribute to a student's well-being. This dichotomy has never been more stark, given the escalating cases of school shootings, which have called for ramped-up physical security measures. Yet, what impact do these changes have on the psychological comfort and mental health of students? In focusing on barriers, alarms, and surveillance, are we unintentionally creating environments of suspicion and anxiety? This chapter aims to probe these complex interrelationships and explore how design can offer holistic solutions. This in-depth examination aligns with the broader theme of this book: the necessity to infuse health and well-being into every aspect of children's spaces.

Understanding Physical Safety in Schools

Physical safety in the context of schools refers to the protection of students, staff, and visitors from tangible harm that could be inflicted by various factors such as intruders, natural disasters, or even everyday accidents like slips and falls. When parents send their children to school, there is an implicit trust that the school's environment is fortified against such threats. This is often made possible through visible safety measures like CCTV cameras, locked doors, and monitored entrances, and security guards..

FIGURE 151. *Learning spaces designed with glass partitions or large windows enable visual permeability, allowing staff and students to naturally survey surrounding areas.*

In discussions around child safety, the narrative often disproportionately skews towards gun violence in schools, whereas there are many threats to student safety and well-being both in and outside of schools. And while the emotional and psychological impact of school shootings are unquestionably devastating, it's crucial to situate these occurrences within the broader landscape of threats to children's physical well-being. The table below offers a comparative look at various physical threats that children face, ranked by the likelihood of their occurrence from most to least. Astonishingly, the data reveals that gun violence in schools, while highly publicized, ranks lower in terms of fatality numbers when compared to issues like motor vehicle accidents, and child abuse. This discrepancy calls for a more nuanced understanding of safety in educational spaces, one that addresses the spectrum of risks while balancing both physical and psychological aspects of child well-being.

TABLE 1. Type of Physical Threat Ranked by Frequency of Occurrences from Most Likely to Occur to Least Likely to Occur:

№	Type of Physical Threat
1	Motor Vehicle Accidents
2	Child Abuse/Neglect
3	Suicide
4	Substance Abuse
5	Bullying and Assault
6	Natural Disasters
7	Environmental Factors
8	Food Allergies and Asthma
9	Gun Violence in Schools

Understanding Psychological Safety in Schools

Psychological safety refers to the emotional and mental well-being that allows students to feel secure, respected, and comfortable in being themselves. Unlike physical safety, which is often overt and observable, psychological safety is subtle and intangible. It encompasses factors like a school's cultural climate, the quality of social interactions, and the design of educational spaces that could either nurture or impair a child's sense of self and emotional well-being.

Psychological safety is not a mere "nice-to-have"; it is a vital component of an educational environment that aspires for comprehensive student well-being. When students feel psychologically safe, they are more likely to engage in creative problem-solving, participate in class discussions, and contribute to a collaborative learning environment. This sense of safety positively correlates with mental health, which in turn influences academic performance. In the absence of psychological safety, children may suffer from anxiety, depression, and other mental health issues that can have long-term adverse effects on their life trajectory.

11 CHAPTER SAFETY

FIGURE 152. Home-like learning environments lead to improved quality of social interactions between students. This improves engagement and potential for collaboration-based pedagogy.

Physical Vs Psychological Safety - A Comparative Analysis

At a fundamental level, physical and psychological safety differ in their goals, methods, and metrics. Physical safety aims to protect from tangible, external threats through concrete measures like surveillance cameras, security personnel, and stringent access controls. Psychological safety, on the other hand, seeks to create an environment where students feel emotionally secure and supported, which is often facilitated through less visible means such as fostering positive relationships, and promoting inclusivity, both of which can be implemented through well-thought-out design features. Metrics for physical safety are often straightforward—number of incidents, for example—while those for psychological safety are more nuanced, often requiring qualitative assessments like surveys or interviews to gauge students' emotional well-being.

11 CHAPTER SAFETY

Physical and psychological safety are not mutually exclusive; rather, they are two sides of the same coin. Both forms of safety contribute to the overall health and well-being of students and staff. *A school fortified like a fortress might be effective in preventing physical harm but can inadvertently create an environment of fear and anxiety.* Conversely, a school that emphasizes psychological comfort but ignores physical safety measures might expose students to external threats. The key is to find a balance that ensures both physical integrity and psychological well-being, allowing students to be safe while they thrive academically and emotionally.

The Ignored Paradigm of Psychological Safety in School Design

FIGURE 153. *A safe school doesn't need to look like a fortress, it can be a well-lit and colorful environment which is furnished with multiple different types of hard and soft seating to learn in variably sized groups or individually. These dynamic spaces are secure but also emotionally comforting.*

When psychological safety is overlooked in the design of educational spaces, the ramifications can be significant and far-reaching. Numerous studies indicate that a lack of psychological safety can lead to heightened stress, anxiety, and depression in students. These emotional states not only impact academic performance but also contribute to long-term mental health issues. In extreme cases, an emotionally unsafe environment can lead to increased rates of truancy, substance abuse, and even self-harm among students.

Due to the increased publicity of gun violence, schools are increasingly becoming fortresses to the point where there is no correlation between the threat and the response. Focused purely on perceived physical danger that far exceeds reality, schools are taking measures that compromise the psychological well-being of children. Metal detectors, security cameras at every corner, and armed guards contribute to a siege mentality, often making students feel like they are under constant threat. This environment is hardly conducive to learning and can be especially counterproductive for children who already come from difficult backgrounds, further exacerbating their stress and anxiety.

The Double-Edged Sword of Safety Measures

While security measures like CCTV, metal detectors, and lockdown drills may bolster physical safety, they can inadvertently contribute to a climate of fear and anxiety. Students may begin to view school as a high-risk environment, a place where danger is always imminent. Such perceptions can have a detrimental impact on their mental health, hindering academic performance and social growth.

Striking a balance between physical and psychological safety involves a thoughtful approach to design. One strategy is to create "home-like" environments with uncluttered spaces and soft seating to foster a sense of comfort. Decentralizing toilet blocks and attaching them to primary learning spaces can reduce the fear and instances of bullying, enhancing psychological safety by eliminating isolated areas. Wayfinding strategies, the use of natural barriers and landscaping instead of imposing fences and employing natural surveillance through glass and good lighting can all contribute to an atmosphere of safety without being overwhelming. These considerations not only adhere to the principles of CPTED (Crime Prevention Through Environmental Design) but also promote a sense of ownership and belonging among students, further contributing to psychological well-being.

Striking the Right Balance

Detailed Design Considerations for Psychological Safety in Schools: Incorporating psychological safety into school design while also adhering to CPTED principles involves a multi-faceted approach. A school environment should not only be secure but also emotionally comforting. Here are some considerations for psychological safety in schools:

1. **Access Control:** Controlling access to the school premises is one of the key CPTED (Crime Prevention Through Environmental Design) principles, especially when it comes to regulating how visitors and parents gain access to student-occupied spaces. Smart access control systems that are intuitive and non-intimidating can enhance both physical and psychological safety. These systems should be designed to minimize bottlenecks during entry and exit, using technology to offer multiple layers of security without being obtrusive. Secure, yet unobtrusive, access points can help streamline the flow of students and staff, reducing congestion and anxiety. One example of this can be found at the building's main entry points: a secure entry sequence allows the school to control different levels of building access, distinguishing between visitors who have access to students and those who have access only to adults. (See diagram below.) For instance, most visitors can meet teachers in the main office conference room, rather than having access to the entire school population of students.

11 CHAPTER SAFETY

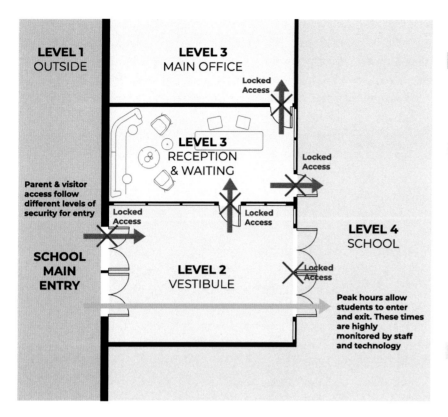

FIGURE 154. Levels of Access Control.

2. **Home-like Environments:** Spaces can be designed to feel more like home, using warm colors, uncluttered layouts, and soft furnishings. This fosters a relaxed, inclusive environment where children feel 'at home,' improving not just comfort but also focus and academic performance.

3. **Decentralized Toilets:** Traditional designs often feature large, centralized toilet blocks, isolated from main activity areas. This can become a hotspot for bullying and induce fear. Decentralizing these facilities and incorporating them closer to primary learning spaces or communal areas improves natural surveillance and accessibility, reduces isolated spots, and lessens the chance of misconduct.

4. **Natural Surveillance:** Learning space design with glass partitions or large windows enable visual permeability, allowing staff and students to naturally survey surrounding areas. This visibility can discourage negative behavior while also creating an open feel that counterbalances the restricted, confined sensation that excessive security measures might produce.

5. **Territorial Reinforcement:** The design can foster a sense of ownership among students by having clearly marked zones, perhaps even co-designed with the students themselves. This makes students more invested in their school spaces and can deter potential offenders.

6. **Wayfinding and Legibility:** Clear signage, wellmarked paths, and the logical organization of spaces improve wayfinding, which is both a CPTED CPTED / Homeland Security principle and a contributor to psychological safety. Students feel more secure when they can easily navigate their environment.

7. **Natural Barriers and Landscaping:** Instead of using intimidating fences and walls, natural elements like hedges, trees, or low-rise earth mounds can define boundaries. These soften the appearance of the school's periphery, reducing the 'fortress' feel while still delineating the school grounds.

8. **Lighting:** Well-lit environments cannot only deter crime but also reduce fear and anxiety. Lighting should be uniform, avoiding dark spots, but it should also be warm and welcoming, rather than stark and institutional, to enhance psychological comfort.

9. **Soft Seating and Safe Zones:** Small alcoves or pockets equipped with soft seating can serve as 'safe zones' for students to retreat, relax, or socialize. These should be clearly visible but subtly integrated into common areas, facilitating both natural surveillance and a feeling of psychological safety.

10. **Auditory Connections:** Having open designs or background noise systems can ensure that children never feel completely isolated, which is crucial in emergency situations and also adds a layer of psychological comfort.

FIGURE 155. Small alcoves or pockets equipped with soft seating can serve as 'safe zones' for students.

Conclusion

The architecture and design of school environments play a significant role in both the physical and psychological well-being of students. While the necessity for physical safety measures cannot be overstated, this chapter supports a balanced approach that also prioritizes psychological safety.

We have unpacked the definitions and importance of both physical and psychological safety, illustrating how they differ yet are deeply interconnected. An overemphasis on only physical security measures can often induce anxiety, stress, and a feeling of confinement among students. On the other hand, design considerations rooted in CPTED and Homeland Security recommendations and psychological safety can work synergistically to create a space that feels safe and welcoming to students.

There are real-world implications of leaning too heavily toward either end of the physical-psychological safety spectrum. Each points to the same truth: A well-rounded, thoughtful approach to design can mitigate risks, enhance well-being, and contribute to more effective learning environments.

Therefore, designers, educators, and policymakers must work collaboratively to build educational spaces that are comprehensive in their approach to safety and well-being. The investment is not just in infrastructure but in the future of the children who will grow within these walls. Let us move beyond traditional design paradigms and aspire to create schools that are truly sanctuaries of learning, growth, and overall well-being.

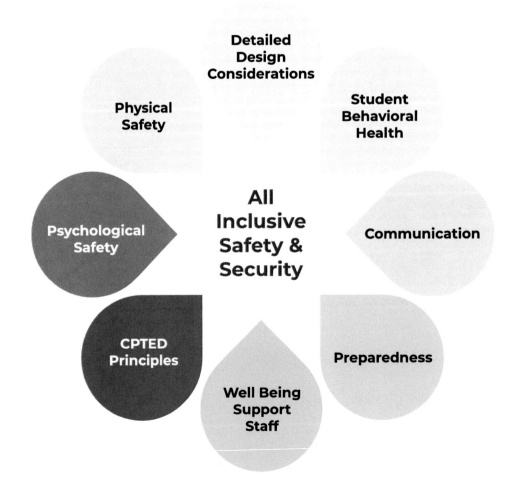

References:

1. Cozens, P., & Love, T. (2015). A review and current status of Crime Prevention Through Environmental Design (CPTED). Journal of Planning Literature, 30(4), 393-412.

2. Cornell, D., & Mayer, M. J. (2010). Why do school order and safety matter? Educational Researcher, 39(1), 7-15.

3. Davidson, J., & Martellozzo, E. (2004). Kids Online: Opportunities and risks for children. Policy Press.

4. Devine, J. (1996). Maximum security: The culture of violence in inner-city schools. University of Chicago Press.

5. Doll, B., Pfohl, W., & Yoon, J. (2010). Handbook of youth prevention science. Routledge.

6. Greenberg, M. T., Weissberg, R. P., O'Brien, M. U., Zins, J. E., Fredericks, L., Resnik, H., & Elias, M. J. (2003). Enhancing school-based prevention and youth development through coordinated social, emotional, and academic learning. American Psychologist, 58(6-7), 466.

7. Leung, R., & Ferris, J. (2018). School environments and social belonging: A meta-analysis. Learning Environments Research, 21(2), 153-175.

8. Newman, O. (1973). Defensible Space: People and Design in the Violent City. Architectural Press.

9. Schargel, F. P., & Priscilla, L. (2009). Bullying prevention for schools: A step-by-step guide to implementing a successful anti-bullying program. Jossey-Bass.

10. Steeves, V., & Webster, C. (2008). Closing the barn door: The effect of parental supervision on Canadian children's online privacy. Bulletin of Science, Technology & Society, 28(1), 4-19.

11. Thapa, A., Cohen, J., Guffey, S., & Higgins-D'Alessandro, A. (2013). A review of school climate research. Review of Educational Research, 83(3), 357-385.

12. U.S. Department of Education, National Center for Education Statistics. (2019). Indicators of School Crime and Safety: 2019 (NCES 2019-047).

11 CHAPTER SAFETY

Prakash Nair, AIA
Bipin Bhadran

Strategies for Successful Implementation[1]

How Should Schools Begin?

Building upon the rationale for and successful examples of schools that are addressing modern education goals using innovative facilities design principles, this chapter answers two questions: 1) What steps can schools take to ensure these practices are implemented to create effective schools? 2) How do schools get started? It provides specific strategies for aligning school buildings with educational aspirations. In the chapter, tools are provided that will assist school leaders in navigating the journey to transform their schools. One such tool describes key indicators comparing traditional and modern elements. Another set of tools equips school leaders to assess the readiness of their schools for such transformational change.

The Change Journey

Leading the effort to transform schools into modern learning environments that will stand the test of time requires a clear vision of the destination. Sharpening the contrast between traditional and new ways of doing business is critical[i]. The distinctions between yesterday's and tomorrow's schools are illustrated in Table 2 and are elaborated upon with examples. Your change journey starts with this list. Compare it with what is actually happening in your school to understand where you are relative to where you want to be. One way to do this is to work closely with the school to create an "Integrated Educational Ecosystem" that represents the aspirational vision of the school (Table 1.)

TABLE 1. Sample Integrated Education Ecosystem to be created working closely with the school's leadership team.

Principles	Outcomes	Methods	Systems	Environment	Assessment
1. Learning is founded in inquiry	1. Thought Leaders and Pioneers in Advanced Technology	1. Shared teacher and student leadership	1. Cohesive instructional goals	1. Fluidly connected learning communities	1. Formative AND summative
2. Learning fosters a culture of curiosity and risk taking	2. Self-directed Learners and Entrepreneurs	2. Regular, cohesive unit planning	2. Flexible teaching arrangements	2. Varied spaces supporting differentiation	2. Continuous and data-driven
3. Mastery of learning is demonstrated in multiple ways	3. Stewards of the Global Environment	3. Multi-cohort instruction	3. Co-facilitated cohort scheduling	3. Spaces for collaborative, individual and active learning	3. Student-constructed
4. Learning is a social process	4. Expert Communicators and Storytellers	4. Thematic integrated projects and courses	4. Need-based instructional grouping (across cohort or subject)	4. Connected to nature	4. Technology-supported
5. Students solve real problems in their local or global community	5. Attuned to people and cultures around them	5. Personalized, technology enhanced learning	5. Multiple credit pathways	5. Supports connection to local & global community	5. Teacher and peer-connected
6. Learning is personalized and learner led					6. Student self-assessment and reflection

[1] A version of this article was originally published in Blueprint for Tomorrow: Redesigning Schools for Student-Centered Learning by Prakash Nair. Harvard Education Press 2018.

CONCLUSION

TABLE 2. Adapted from 21st CenturySchools.com[ii].

Learning and Space Design Implications of Traditional vs. Modern Schools	
Yesterday's Schools	**Tomorrow's Schools**
Time bound: Instruction takes place within designated periods	Outcome bound: Learning is the constant driver with time being flexible
Assessment focused on memorization of facts	Assessment focused on what students know and can do
Instruction focused on lower levels of Bloom's Taxonomy: Remembering	Instruction focused on higher levels of Bloom's Taxonomy: Evaluating and Creating
Textbook-based curriculum	Research-based curriculum
Passive, teacher-directed learning	Active, student-directed learning
Learners isolated in classrooms working individually or in small groups	Learners working collaboratively and with peers around the world
Classrooms arranged primarily for direct instruction	Classrooms arranged for multiple forms of learning
Fragmented curriculum taught in isolation by single content	Interdisciplinary curriculum taught with cross-content connections
Strong focus on theory	Strong focus on practice
Teachers work in isolation	Teachers collaborate with colleagues within and beyond their school
Print serves as the primary vehicle for learning and assessment	Learning is focused on performance, projects, and portfolio-based assessments
Centralized classroom technology	Disbursed, personalized technology
Inadequate outdoor learning opportunities	Outdoor spaces designed for informal learning experiences
Primary focus on efficiency	Primary focus on effectiveness

Learners isolated in classrooms working individually or in small groups versus Learners working collaboratively and with peers around the world

Learners in a traditional classroom work in isolation, most often on individual assignments as the focus, while learning in a modern school involves collaboration with their peers who may be located in their school, in other schools within the district and often with students of similar interests around the world. Physically students in the vast majority of present day schools are confined in a box called a classroom where the assumption is that most of the learning will take place. In the learning environments of tomorrow, students collaborate with each other, but also with global peers. Under this model, the school is not so important as the center of learning, but rather as a gateway to a whole world of learning opportunities – through connections with people and resources all over the world.

CONCLUSION

FIGURE 156. *The DNA of most school buildings starts with teacher-centered learning and the classroom is perfectly suited for that.*

FIGURE 157. *The shift to a more student-centered model of education is easier to do when the spaces themselves are conducive to that model.*

CONCLUSION

FIGURE 158. Classroom settings are naturally limiting in terms of the different modalities of learning they permit. They are primarily designed for direct instruction.

FIGURE 159. Freed from the Classroom box, students are free to learn in different ways both individually and in teams with greater opportunities for autonomy.

Classrooms arranged primarily for direct instruction versus Learning Studios arranged for multiple forms of learning

Learning in traditional schools involves mostly direct instruction, drill and practice, whereas schools that recognize the importance of autonomy, collaboration, and hands-on experiences support multiple forms of learning. Direct instruction is just one of the many forms. For example, the design of a "learning studio" must be very different from a traditional classroom because it needs to support self-directed, collaborative, and experiential learning modalities.

There are big gaps that exist in education – between research and action, between stated goals and policy, and between perception and reality. The contrasting Learning and Space Design Implications of yesterday's and tomorrow's schools illuminate many of these gaps. In education, though research unequivocally supports a student-centered model, schools and school systems overwhelmingly practice the traditional mass-production model of schooling. Breaking away from the industrial model is a moral imperative if students are to be competitive in a global economy.

Getting Started

The Discovery Process: Regardless of whether you have $100,000 or $100 million to spend on your school facilities, it is imperative that you begin with a rigorous "discovery" process which contains several discrete steps that need to be followed in order to ensure that you are able to achieve meaningful, sustainable change. For smaller projects within individual schools, you can go through this process on your own but for bigger projects and those that involve multiple schools, we strongly recommend that you retain the services of an educational facility planner and design professional.

Establish a Leadership Team: Individual superintendents and school principals tend to relegate matters related to school facilities to their in-house engineering professionals. This is a mistake. Facilities people are great when it comes to managing the day-to-day affairs of the school physical plant but they have little authority to make the kinds of reform-oriented decisions that need to be made for true educational transformation to happen. This results in the expenditure of billions of dollars in capital spending each year across the globe for which we see very little in the way of educational results. That is why we recommend a school project should begin with the appointment of a Leadership Team whose size and composition would vary depending upon the scope and scale of the project.

Generally, for district-wide master plans, large renovations, school additions or new school projects a Leadership Team should be large enough to include as many key stakeholders as possible but small enough to be effective. We have worked on projects where Leadership Teams had representatives from the district and/or school management (like the Superintendent or Principal) plus respected teachers, parents, students, community representatives and local business leaders. Ideally, Leadership Teams should not exceed 12-15 people. To avoid the complications of creating a new legal entity, we suggest that Leadership Teams should only have the authority to make recommendations. The legally responsible school executives and school boards can use these to craft their own decisions. While Leadership Teams may not have any official standing, their work will carry weight with the community because they take into consideration the opinions and wishes of all education stakeholders.

Hire Your Professional Team Early: Avoid the temptation to define your project too early in the process. Often, school leaders will have a strong idea about what they want and then go through the motions to post-rationalize their decisions. In fact, most school facilities projects in the US are already fully defined in terms of scope and budget before facilities professionals have even been hired. The better approach is to write a very generic scope of work for the facilities professional that talks generally about an individual school or district's aspirations and contains a brief description of the project and budget which can be used to solicit proposals from interested teams of school facility planners and architects.

CONCLUSION

FIGURE 160. *Ideally, leadership teams should be around 12-15 people and be a representative cross section of your school community.*

Be mindful that schools and school districts make few decisions that have more long-term influence over the lives of thousands of students than their decision to hire a facilities planner/architect. That is why it is imperative to make this decision carefully and thoughtfully. Your selected design professional will lead you through the remaining steps in the Discovery Process below including the following workshops:

Leading Practices Workshop: This will be the event that officially kicks-off your facility planning project. The purpose of this workshop is twofold. First, it will introduce your community to the current leading practices in education and school facilities from around the country and the world. It will contain relevant, successful case studies and offer data-driven evidence that new models of education, supported modern school facilities do work. This is important because it will help participants change their frame of reference from what they know about education and school buildings (which may be based on their own memories of school) to the current state of educational innovation and research. Second, it will offer all participants an opportunity to be personally and directly involved in important decisions that will shape their community's educational future. This and the other workshops described below will all help build consensus for the solutions that will emerge from the Discovery process.

Invite all key stakeholders to attend this workshop including students, teachers, parents, local community reps, school district officials, local press and any others who are active in education in your community. Hold this meeting after work hours – from 6:30 pm to 9:00 pm is usually a good time. If possible, offer childcare for parents with very young children to increase participation. The workshop can follow one of two formats. It can either be led by your planning and design professional or it can be set up as a panel discussion with local leaders from different fields in attendance.

The workshop should include time for the invited professionals to be heard and after that attendees will work in teams to ask questions, contribute their ideas and express their concerns. It is a good idea to provide each attendee with post-it notes on which they can jot down their ideas and concerns and place on large sheets of paper that have been affixed to the walls around the room. After all the ideas are posted, each attendee could be given five green stickers which they can use to vote for the ideas around the room. The organizers of the event will collect and collate the information gathered from the workshop.

If possible, have the event be covered by local TV stations and produce an in-house video that can be shared on the school and/or district website.

Visioning Workshop: This workshop of about 90 minutes duration will be for the Leadership Team. It should

CONCLUSION

preferably be held soon after the Leading Practices session. During this workshop, some of the leading practices material will be reviewed as well as feedback from the larger community meeting. The purpose of this workshop is to have a clear vision for the project moving forward. This team will also study the differences between traditional and modern schools (discussed earlier in this chapter) to understand where their school fits in and get a sense of the gaps that need to be filled to achieve their educational vision.

Focus Groups: We recommend that three focus groups be conducted for teachers, parents, and students, each lasting about one hour. The purpose of these focus groups is to understand the areas of consensus and disagreement between these key stakeholder groups. Similar questions would be posed to each group and the collated results will identify areas of concern. For example, one question might ask to rank the following in order of importance: Lecture, Team Collaboration, Independent Study and Hands-on Work. Should teachers rank lecture as most important, parents prioritize team collaboration and students prefer hands-on work, it gives the planner a basis upon which to start a productive dialogue with each group and then with the school leadership about the best way to design spaces that meet the needs of each stakeholder group. Like the other workshops, focus groups will provide important data but the process itself is intended to be about people and getting them involved and excited to be a part of the transformation journey.

Assessment of Existing Spaces: In the book, Blueprint for Tomorrow published by Harvard Education Press that this chapter is extracted from, we have provided three checklists (one for early childhood facilities, one for elementary school facilities and one for secondary school facilities) which is a hybrid tool based on Dr. Lorraine Maxwell's Classroom Assessment Scale . The Leadership Team should use these checklists for a quick and accurate assessment of their school facility's effectiveness as a place to deliver a modern education. Where multiple schools are involved, then at a minimum, a representative sample of schools should be evaluated. These checklists are also valuable because they provide educators and laypersons on the team with a tool to measure the quality of the designs that emerge. Naturally, every effort must be made to ensure that the selected design scores as close to the maximum on the list as possible. We have developed new assessment tools that are easier to use as mobile APPS. These tools are valuable not only to assess the educational effectiveness of existing buildings and new building designs, but they can also be used to monitor the way new and renovated school buildings are actually being used. The following three APPS are examples of the kinds of mobile tools that are essential to include as part of any sustainable change process.

FIGURE 161. High quality learning spaces are just the beginning. Their effectiveness will depend on how they inspire the important work of building a new curriculum, realigning schedules, choosing appropriate resources including technologies and educational partners and, most important, professional development designed to help teachers get the most out of their new or renovated spaces. Lauren Mehrbach, learning designer and head of middle school at Singapore American school, running a facilities related PD workshop for teachers at American Embassy School in New Delhi, India.

EduSPACE
Dynamic Learning Maps

WHAT IS IT? EduSPACE is the ONLY APP in the world that **dynamically and continuously** measures the quality of teaching and learning spaces at a school.

HOW DOES IT WORK? Floor plans of new and renovated schools are uploaded via a desktop portal which then show up on the user's mobile APP. Pins mark each space. Short (three-to-five-minute) "audits" are conducted by students and teachers themselves of selected areas that record the actual use and occupancy of each individual space at different times of each day.

WHAT DO SCHOOLS GET? Accurate and reliable data that show:
1. How different spaces in the school are actually used at different time of the day – as opposed to their anticipated use
2. The modalities of learning most prevalent in different areas and in the school at large
3. Areas that are properly utilized vs. those that are under or over-utilized
4. The extent to which newly designed spaces (renovations, additions or new buildings) yield corresponding educational benefits

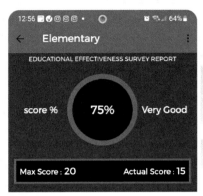

SPACE
Efficacy of Learning Spaces

WHAT IS IT? SPACE is a **fully customizable** APP that records the effectiveness of learning environments to support each school's educational vision and mission.

HOW DOES IT WORK? Following a walk-through of the school and the creation of a photographic essay of all the learning spaces, designated school representatives "assess" how well the building serves stated educational goals. Existing buildings are benchmarked before any renovations are done, the design is assessed for quality for both new and existing buildings and follow-up assessments are done when the spaces are in use.

WHAT DO SCHOOLS GET?
1. Customized templates they create themselves using research-based criteria focused on environmental conditions that improve student health, well-being and achievement
2. Different scores for different age levels – Early Childhood, Elementary, Middle School and High School that accurately record the extent to which learning spaces support teaching and learning
3. The ability to prioritize spending decisions in a way that will yield the greatest educational value for the money spent

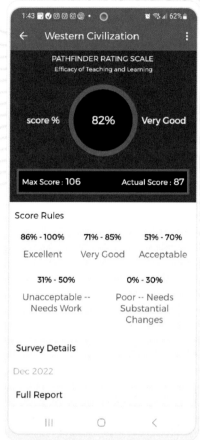

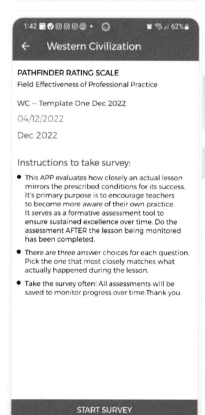

PATH
Connecting Pedagogy & Space

WHAT IS IT? PATH (short for Pathfinder) is a **fully customizable** APP that lets teachers create and measure the effectiveness of lesson plans that maximize the learning potential of innovative learning spaces.

HOW DOES IT WORK? This APP measures how closely an actual lesson mirrors the prescribed conditions for its success. Teachers create a set of pedagogical criteria on a desktop portal which then show up in their APP next to the lesson being assessed. The criteria used to deliver any particular lesson are designed to take full advantage of collaborative, flexible and innovative learning spaces available at the school.

WHAT DO SCHOOLS GET?
1. Customized lesson plans describing the skills and competencies that students will practice within each lesson
2. The efficacy of each lesson or multidisciplinary curriculum unit can be measured on the app in under one minute
3. Encourages teachers to take full advantage of innovative learning spaces
4. The ability to get instant feedback after each lesson makes the APP a highly effective formative assessment tool that teachers will be able to use to improve their own practice

Sustainable Design and Local Ethos Workshop: This workshop for the Leadership Team will explore all the ways in which the school project at hand can be made sustainable or "green". Opportunities to connect sustainable design strategies to the school curriculum should also be part of the agenda so that students will continue to benefit from the green products and technologies that are used well after the construction is complete. Part of sustainable design practice is to connect design and construction to local climate, materials, and practices. It is therefore important to discuss the ethos of the school and establish a signature that is tied to elements that have local significance. Unless this is done consciously, there is no guarantee that the design solutions which emerge will be sustainable or sensitive to the local context.

Design Analysis and Professional Development (Curriculum Mapping) Workshop: On-site Space-based Curriculum Mapping can occur once the design has been developed through the completion of schematics that include furnishing layouts. This is a hands-on workshop with teachers and administrators to understand how the new design can support their professional needs. The activity will help educators to capitalize on better designed spaces to improve current practice. It will also aid them to incorporate new pedagogies, re-arrange learning activities, set up teaming arrangements with other faculty, and utilize the learning environment in innovative ways that the old campus did not permit.

Facility-Related Professional Development: This workshop is sometimes referred to as "Educational Commissioning" and it is offered immediately before school opening. It is designed to give teachers an opportunity to walk the newly renovated or constructed school facility and explore the many ways in which it can enhance teaching and learning. This workshop can also help systematize teaming approaches so that even new teachers who join later can more easily fit into the collaborative framework reinforced by the design of the new facility.

Except for the Educational Commissioning workshop, all the workshops and exercises described above will be conducted before any significant design work has started. These represent a good investment for the school from the perspective of returning a superior design that will have the support of the community. By garnering the input and feedback of key stakeholder groups early, these workshops, which can be held over two days, can also save time and shorten the overall project duration.

Change Management Support: A world-class facility setup for 21st century teaching and learning requires a certain level of preparation that school principals may not have the time, resources or experience to manage. Professionals who are working as part of the project team would assist the Principal on any or all of the following items: 1) Schedule adjustments as appropriate to leverage the new facility plan; 2) Broker teacher-teacher and principal-principal video conference dialogues with existing innovative, schools; 3) Build a collaboration model facilitated by the new facilities that fits in with or enhances current teacher practices; 4) Develop an ongoing professional development strategy based on a gap analysis of where teachers stand today relative to what may be expected of them in the newly designed facility; 5) Design communication strategies to help parents and other key stakeholders understand how kids will be benefit from the new learning environment.

References:

[i] J. Stewart Black and Hal B. Gregersen, Leading strategic change: Breaking through the brain barrier. (Upper Saddle River, NJ: Pearson Education, Inc. 2003).

[ii] 21st Century Schools, "What is 21st Century Education?" http://www.21stCenturySchools.com/What_is_21st_Century_Education.htm

[ii] Lorraine Maxwell, "Competency in Child Care Settings: The Role of the Physical Environment," Environment and Behavior 39, no. 2 (2007): 229-245.

Additional Photo Credits

Several images featured in this book are from the following projects executed by Fielding Nair International (FNI) during Prakash Nair's tenure as its Founding President & CEO. FNI owns exclusive or shared copyright to these images which are not to be reproduced in any form without explicit written consent of the copyright holders. Other members of Education Design International's current team who previously served as Principals or Lead Designers on many of these projects include Director of Design Louis Sirota, Principal Architect Jay Litman and Sr. Interior Designer Danielle McCarthy.

St. Francis of Assisi Catholic Elementary School
Kingston, Ontario Canada

Emerald Elementary School, Boulder Valley, CO

Strathcona Tweedsmuir School, Calgary, Canada

Anne Frank Inspire Academy, San Antonio, TX

Academy of the Holy Names, Tampa, FL

International School of Brussels, Belgium

High School for Recording Arts, St. Paul, MN

International School of Dusseldorf, Germany

Shorecrest Preparatory School, St. Petersburg, FL

Hillel Academy, Tampa, FL

Meadowlark School, Boulder Valley, CO

PK Yonge Developmental Research School, Gainesville, FL

Col.legi Montserrat, Barcelona, Spain

Yew Chung International School of Chongqing, China

Fisher STEAM Midde School, Greenville, SC

Hillel School of Detroit, MI

Creekside Elementary School, Boulder Valley, CO

ABOUT THE AUTHORS

Prakash Nair, AIA
Founding President & CEO, Education Design International (EDI)

Prakash is a futurist and visionary architect who leads Education Design International, a firm that has consulted in 58 countries across six continents. He has received numerous international awards, including the A4LE MacConnell Award, the highest worldwide honor in school design. Prakash has published extensively in leading international journals and is the author of three books, including the landmark publication "Blueprint for Tomorrow: Redesigning Schools for Student-Centered Learning," published by Harvard Education Press.
Contact: Prakash@EducationDesign.com

Parul Minhas, Ph.D.
Director of Research & Digital Innovation (EDI)

Parul Minhas is known for her groundbreaking research in learning environments that foster student health and well-being. She spearheads the development of various EDI APPS with a focus on educational excellence and collaborates with Prakash Nair on influential studies in school design. Her research and keynote presentations on neuroarchitecture and biophilic design have garnered global acclaim.
Contact: Parul@EducationDesign.com

Roni Zimmer Doctori
Principal Architect & Israel Studio Head (EDI)

With over 17 years in architectural expertise, Roni Zimmer Doctori stands as a testament to innovation in educational space design. She's the co-author of the influential book, "Learning by Design: Live | Play | Engage | Create." Having shared her insights from Australia to Canada, her pioneering projects, like the GOGYA Teacher Training Academy, are globally acclaimed. An honored graduate of the Technion – Israel Institute of Technology, Roni's fervor stems from her children's educational experiences. She champions transformative school architectures and vibrant, student-focused curriculums.
Contact: Roni@EducationDesign.com

ABOUT THE AUTHORS

Bipin Bhadran
Managing Director (EDI)

Leading EDI's Bangalore studio, Bipin Bhadran excels in harmonizing designs with global educational trends. With accolades including the nation's top design award, his portfolio shines with projects like the Istanbul International Community School and the World Bank's Central Asian education revamps. Bipin's strength lies in uniting diverse stakeholder visions and ensuring exceptional project value. An MBA from Leeds University Business School, UK, enhances his expertise.
Contact: bipin@educationdesign.com

Louis Sirota, AIA
Director of Design (EDI)

Louis Sirota stands as a force in the realm of educational facility design. He possesses a unique ability to interpret the nuanced needs of schools, translating them into functional and innovative designs. Louis has spearheaded projects like the YCIS Secondary School in Chongqing and Texas Tech's Satellite Campus in Costa Rica. His collaboration with the Yew Chung International Educational Foundation redefined learning spaces across China. With a vision rooted in transformative architecture, his master's thesis at North Dakota State University heralded a new age for academic libraries, envisioning them as dynamic, tech-integrated learning hubs.
Contact: Louis@EducationDesign.com

Gary Stager, Ph.D.
Senior Education & S.T.E.M. Consultant (EDI)

A trailblazer in education technology, Dr. Gary Stager's impact ranges from spearheading 1:1 computing in the '90s to influential collaborations at the MIT Media Lab. Recognized as a "shaper of our future" by Converge Magazine and awarded by CUE, his expertise embraces STEM, the maker movement, and fostering environments where passion for learning thrives. Notably, he collaborated with educational pioneer Seymour Papert and now curates the world's most extensive Papert archive.
Contact : gary@stager.org

ABOUT THE AUTHORS

Anna Harrison
Director, Educational Innovation (EDI)

Anna Harrison is a seasoned expert in biophilic design, focusing on sustainable and regenerative planning at Education Design International. She collaborates with educational leaders to transform entire school campuses into regenerative learning environments, contributing to both social and environmental justice. Anna holds a Master's in Education Leadership and is accredited in Learning Environments Planning (ALEP) and Energy and Environmental Design (LEED AP). She is also an NCIDQ-certified interior designer. Known for her international conference presentations, Anna's work has gained global recognition.
Contact: Anna@EducationDesign.com

Francesco Cupolo
Community Outreach Coordinator (EDI)

Hailing from Caracas and rooted in Philadelphia, Francesco is a devoted educator and designer with a knack for community engagement. With a rich background spanning Industrial Design to Anthropology, he's transformed makerspace education in K-12 settings. Beyond his 8-year tenure in afterschool program coordination, Francesco is renowned for crafting interactive exhibits and creative workshops. A graduate of the Pacific Northwest College of Art's entrepreneurship program, his commitment is evident in fostering creativity across age groups.
Contact: Francesco@EducationDesign.com

Karin Nakano

A Swarthmore College graduate specialized in Educational Studies and Computer Science, Karin Nakano combines her skills to design residential learning facilities that encourage peer-to-peer interaction. With a strong inclination towards biophilic design, she is dedicated to creating educational spaces that are not only functional but also deeply enriching. Passionate about equal access to learning environments that inspire joy and lifelong curiosity, Karin integrates natural elements into her designs to enhance the human experience.
Contact: karin.nakano06@gmail.com

ABOUT THE AUTHORS

Paras Sareen
Founder & CEO (PSA Landscape Architects)

Paras Sareen is the founder of PSA Landscape Architects, specializing in a diverse range of landscape projects including institutional, residential, and urban settings. Known for his sustainable and low-maintenance designs that incorporate native and adaptive vegetation, Paras also contributes his expertise to the academic field. His research explores creative strategies to bring nature into urban educational settings, aiming to foster cognitive growth and environmental stewardship in students.
Contact: Paras@psa.landscape@gmail.com

Jeff Melendez, Ed.D.
Founder and CEO (Zeal Education Group)

Jeff Melendez has 25 years of experience in public education, including six years as a superintendent and 12 years as an Adjunct Professor at Teachers College, Columbia University. He founded Zeal Education Group, a consultancy dedicated to addressing challenges in the education sector which include school safety and cyber-security audits, and reimagining instructional spaces. Jeff holds M.A., Ed.M., and Ed.D. degrees from Teachers College and is certified by the National Board for Professional Teaching Standards.
Contact: jeff@zeal-ed.com

John K. Ramsey, CAE ALEP
CEO of Association for Learning Environments (A4LE)

John Ramsey has nearly three decades of leadership experience in non-profit international associations, including a 16-year tenure at the Association for Learning Environments (A4LE). Under his leadership, A4LE expanded its membership from 2,800 to over 5,800 globally and introduced initiatives like the Advanced Academy for Learning Environment Planning and the Accredited Learning Environment Planner certification program. John previously served as Executive Director for the National Center for State Courts and held multiple roles at the Association of Public Safety Communications Officials. He's a recently named A4LE Fellow and a retired U.S. Army officer.
Contact: john@a4le.org

The Association for Learning Environments (A4LE) is a global non-profit organization dedicated to improving and transforming the places where children learn. Founded in 1921 as the National Council on Schoolhouse Construction and later renamed to reflect its broader mission, A4LE serves thousands of professionals working in various facets of education, design, and construction.

The association believes that well-designed learning environments can greatly impact student success and teacher effectiveness. To that end, A4LE offers professional development, research initiatives, and opportunities for collaboration among architects, planners, educators, and administrators.

By fostering an interdisciplinary dialogue, A4LE aims to create innovative, sustainable, and enriching learning environments that cater to the evolving needs of students and educators around the world.

Made in the USA
Coppell, TX
04 October 2023